Logic: A Very Short Introduction

Praise for the First Edition

'A delightful and engaging introduction to the basic concepts of logic. Whilst not shirking the problems, Priest always manages to keep his discussion accessible and instructive.'

Adrian Moore, St Hugh's College, Oxford

'An excellent way to whet the appetite for logic.... Even if you read no other book on modern logic but this one, you will come away with a deeper and broader grasp of the *raison d'être* for logic.'

Chris Mortensen, University of Adelaide

'This new addition to the VSI series makes a splendid fist of warming up logic for the general palate. You still have to learn a few symbols, but the pleasure in following Priest's mind-bending puzzles about dividing amoebas, or refuting, with his guidance, a famous refutation of the existence of time, will be worth it.'

The Guardian

'This elegant and beautifully written little book provides the novice with an excellent introduction to the general field of logic.'

Michael Detlefsen, University of Notre Dame

VERY SHORT INTRODUCTIONS are for anyone wanting a stimulating and accessible way into a new subject. They are written by experts, and have been translated into more than 45 different languages.

The series began in 1995, and now covers a wide variety of topics in every discipline. The VSI library now contains over 500 volumes—a Very Short Introduction to everything from Psychology and Philosophy of Science to American History and Relativity—and continues to grow in every subject area.

Very Short Introductions available now:

ACCOUNTING Christopher Nobes
ADOLESCENCE Peter K. Smith
ADVERTISING Winston Fletcher
AFRICAN AMERICAN RELIGION Eddie S. Glaude Jr
AFRICAN HISTORY John Parker and Richard Rathbone
AFRICAN RELIGIONS Jacob K. Olupona
AGEING Nancy A. Pachana
AGNOSTICISM Robin Le Poidevin
AGRICULTURE Paul Brassley and Richard Soffe
ALEXANDER THE GREAT Hugh Bowden
ALGEBRA Peter M. Higgins
AMERICAN HISTORY Paul S. Boyer
AMERICAN IMMIGRATION David A. Gerber
AMERICAN LEGAL HISTORY G. Edward White
AMERICAN POLITICAL HISTORY Donald Critchlow
AMERICAN POLITICAL PARTIES AND ELECTIONS L. Sandy Maisel
AMERICAN POLITICS Richard M. Valelly
THE AMERICAN PRESIDENCY Charles O. Jones
THE AMERICAN REVOLUTION Robert J. Allison
AMERICAN SLAVERY Heather Andrea Williams
THE AMERICAN WEST Stephen Aron
AMERICAN WOMEN'S HISTORY Susan Ware
ANAESTHESIA Aidan O'Donnell
ANARCHISM Colin Ward
ANCIENT ASSYRIA Karen Radner
ANCIENT EGYPT Ian Shaw
ANCIENT EGYPTIAN ART AND ARCHITECTURE Christina Riggs
ANCIENT GREECE Paul Cartledge
THE ANCIENT NEAR EAST Amanda H. Podany
ANCIENT PHILOSOPHY Julia Annas
ANCIENT WARFARE Harry Sidebottom
ANGELS David Albert Jones
ANGLICANISM Mark Chapman
THE ANGLO-SAXON AGE John Blair
ANIMAL BEHAVIOUR Tristram D. Wyatt
THE ANIMAL KINGDOM Peter Holland
ANIMAL RIGHTS David DeGrazia
THE ANTARCTIC Klaus Dodds
ANTISEMITISM Steven Beller
ANXIETY Daniel Freeman and Jason Freeman
THE APOCRYPHAL GOSPELS Paul Foster
ARCHAEOLOGY Paul Bahn
ARCHITECTURE Andrew Ballantyne
ARISTOCRACY William Doyle
ARISTOTLE Jonathan Barnes
ART HISTORY Dana Arnold
ART THEORY Cynthia Freeland
ASIAN AMERICAN HISTORY Madeline Y. Hsu
ASTROBIOLOGY David C. Catling
ASTROPHYSICS James Binney

ATHEISM Julian Baggini
THE ATMOSPHERE Paul I. Palmer
AUGUSTINE Henry Chadwick
AUSTRALIA Kenneth Morgan
AUTISM Uta Frith
THE AVANT GARDE David Cottington
THE AZTECS Davíd Carrasco
BABYLONIA Trevor Bryce
BACTERIA Sebastian G. B. Amyes
BANKING John Goddard and John O. S. Wilson
BARTHES Jonathan Culler
THE BEATS David Sterritt
BEAUTY Roger Scruton
BEHAVIOURAL ECONOMICS Michelle Baddeley
BESTSELLERS John Sutherland
THE BIBLE John Riches
BIBLICAL ARCHAEOLOGY Eric H. Cline
BIOGRAPHY Hermione Lee
BLACK HOLES Katherine Blundell
BLOOD Chris Cooper
THE BLUES Elijah Wald
THE BODY Chris Shilling
THE BOOK OF MORMON Terryl Givens
BORDERS Alexander C. Diener and Joshua Hagen
THE BRAIN Michael O'Shea
BRANDING Robert Jones
THE BRICS Andrew F. Cooper
THE BRITISH CONSTITUTION Martin Loughlin
THE BRITISH EMPIRE Ashley Jackson
BRITISH POLITICS Anthony Wright
BUDDHA Michael Carrithers
BUDDHISM Damien Keown
BUDDHIST ETHICS Damien Keown
BYZANTIUM Peter Sarris
CALVINISM Jon Balserak
CANCER Nicholas James
CAPITALISM James Fulcher
CATHOLICISM Gerald O'Collins
CAUSATION Stephen Mumford and Rani Lill Anjum
THE CELL Terence Allen and Graham Cowling
THE CELTS Barry Cunliffe
CHAOS Leonard Smith
CHEMISTRY Peter Atkins
CHILD PSYCHOLOGY Usha Goswami
CHILDREN'S LITERATURE Kimberley Reynolds
CHINESE LITERATURE Sabina Knight
CHOICE THEORY Michael Allingham
CHRISTIAN ART Beth Williamson
CHRISTIAN ETHICS D. Stephen Long
CHRISTIANITY Linda Woodhead
CIRCADIAN RHYTHMS Russell Foster and Leon Kreitzman
CITIZENSHIP Richard Bellamy
CIVIL ENGINEERING David Muir Wood
CLASSICAL LITERATURE William Allan
CLASSICAL MYTHOLOGY Helen Morales
CLASSICS Mary Beard and John Henderson
CLAUSEWITZ Michael Howard
CLIMATE Mark Maslin
CLIMATE CHANGE Mark Maslin
CLINICAL PSYCHOLOGY Susan Llewelyn and Katie Aafjes-van Doorn
COGNITIVE NEUROSCIENCE Richard Passingham
THE COLD WAR Robert McMahon
COLONIAL AMERICA Alan Taylor
COLONIAL LATIN AMERICAN LITERATURE Rolena Adorno
COMBINATORICS Robin Wilson
COMEDY Matthew Bevis
COMMUNISM Leslie Holmes
COMPLEXITY John H. Holland
THE COMPUTER Darrel Ince
COMPUTER SCIENCE Subrata Dasgupta
CONFUCIANISM Daniel K. Gardner
THE CONQUISTADORS Matthew Restall and Felipe Fernández-Armesto
CONSCIENCE Paul Strohm
CONSCIOUSNESS Susan Blackmore
CONTEMPORARY ART Julian Stallabrass
CONTEMPORARY FICTION Robert Eaglestone
CONTINENTAL PHILOSOPHY Simon Critchley
COPERNICUS Owen Gingerich
CORAL REEFS Charles Sheppard
CORPORATE SOCIAL RESPONSIBILITY Jeremy Moon

CORRUPTION Leslie Holmes
COSMOLOGY Peter Coles
CRIME FICTION Richard Bradford
CRIMINAL JUSTICE Julian V. Roberts
CRITICAL THEORY Stephen Eric Bronner
THE CRUSADES Christopher Tyerman
CRYPTOGRAPHY Fred Piper and Sean Murphy
CRYSTALLOGRAPHY A. M. Glazer
THE CULTURAL REVOLUTION Richard Curt Kraus
DADA AND SURREALISM David Hopkins
DANTE Peter Hainsworth and David Robey
DARWIN Jonathan Howard
THE DEAD SEA SCROLLS Timothy H. Lim
DECOLONIZATION Dane Kennedy
DEMOCRACY Bernard Crick
DEPRESSION Jan Scott and Mary Jane Tacchi
DERRIDA Simon Glendinning
DESCARTES Tom Sorell
DESERTS Nick Middleton
DESIGN John Heskett
DEVELOPMENTAL BIOLOGY Lewis Wolpert
THE DEVIL Darren Oldridge
DIASPORA Kevin Kenny
DICTIONARIES Lynda Mugglestone
DINOSAURS David Norman
DIPLOMACY Joseph M. Siracusa
DOCUMENTARY FILM Patricia Aufderheide
DREAMING J. Allan Hobson
DRUGS Les Iversen
DRUIDS Barry Cunliffe
EARLY MUSIC Thomas Forrest Kelly
THE EARTH Martin Redfern
EARTH SYSTEM SCIENCE Tim Lenton
ECONOMICS Partha Dasgupta
EDUCATION Gary Thomas
EGYPTIAN MYTH Geraldine Pinch
EIGHTEENTH-CENTURY BRITAIN Paul Langford
THE ELEMENTS Philip Ball
EMOTION Dylan Evans
EMPIRE Stephen Howe
ENGELS Terrell Carver
ENGINEERING David Blockley
ENGLISH LITERATURE Jonathan Bate
THE ENLIGHTENMENT John Robertson
ENTREPRENEURSHIP Paul Westhead and Mike Wright
ENVIRONMENTAL ECONOMICS Stephen Smith
ENVIRONMENTAL LAW Elizabeth Fisher
ENVIRONMENTAL POLITICS Andrew Dobson
EPICUREANISM Catherine Wilson
EPIDEMIOLOGY Rodolfo Saracci
ETHICS Simon Blackburn
ETHNOMUSICOLOGY Timothy Rice
THE ETRUSCANS Christopher Smith
EUGENICS Philippa Levine
THE EUROPEAN UNION John Pinder and Simon Usherwood
EUROPEAN UNION LAW Anthony Arnull
EVOLUTION Brian and Deborah Charlesworth
EXISTENTIALISM Thomas Flynn
EXPLORATION Stewart A. Weaver
THE EYE Michael Land
FAMILY LAW Jonathan Herring
FASCISM Kevin Passmore
FASHION Rebecca Arnold
FEMINISM Margaret Walters
FILM Michael Wood
FILM MUSIC Kathryn Kalinak
THE FIRST WORLD WAR Michael Howard
FOLK MUSIC Mark Slobin
FOOD John Krebs
FORENSIC PSYCHOLOGY David Canter
FORENSIC SCIENCE Jim Fraser
FORESTS Jaboury Ghazoul
FOSSILS Keith Thomson
FOUCAULT Gary Gutting
THE FOUNDING FATHERS R. B. Bernstein
FRACTALS Kenneth Falconer
FREE SPEECH Nigel Warburton
FREE WILL Thomas Pink
FREEMASONRY Andreas Önnerfors
FRENCH LITERATURE John D. Lyons

THE FRENCH REVOLUTION William Doyle
FREUD Anthony Storr
FUNDAMENTALISM Malise Ruthven
FUNGI Nicholas P. Money
THE FUTURE Jennifer M. Gidley
GALAXIES John Gribbin
GALILEO Stillman Drake
GAME THEORY Ken Binmore
GANDHI Bhikhu Parekh
GENES Jonathan Slack
GENIUS Andrew Robinson
GEOGRAPHY John Matthews and David Herbert
GEOPOLITICS Klaus Dodds
GERMAN LITERATURE Nicholas Boyle
GERMAN PHILOSOPHY Andrew Bowie
GLOBAL CATASTROPHES Bill McGuire
GLOBAL ECONOMIC HISTORY Robert C. Allen
GLOBALIZATION Manfred Steger
GOD John Bowker
GOETHE Ritchie Robertson
THE GOTHIC Nick Groom
GOVERNANCE Mark Bevir
GRAVITY Timothy Clifton
THE GREAT DEPRESSION AND THE NEW DEAL Eric Rauchway
HABERMAS James Gordon Finlayson
THE HABSBURG EMPIRE Martyn Rady
HAPPINESS Daniel M. Haybron
THE HARLEM RENAISSANCE Cheryl A. Wall
THE HEBREW BIBLE AS LITERATURE Tod Linafelt
HEGEL Peter Singer
HEIDEGGER Michael Inwood
HEREDITY John Waller
HERMENEUTICS Jens Zimmermann
HERODOTUS Jennifer T. Roberts
HIEROGLYPHS Penelope Wilson
HINDUISM Kim Knott
HISTORY John H. Arnold
THE HISTORY OF ASTRONOMY Michael Hoskin
THE HISTORY OF CHEMISTRY William H. Brock
THE HISTORY OF LIFE Michael Benton
THE HISTORY OF MATHEMATICS Jacqueline Stedall
THE HISTORY OF MEDICINE William Bynum
THE HISTORY OF TIME Leofranc Holford-Strevens
HIV AND AIDS Alan Whiteside
HOBBES Richard Tuck
HOLLYWOOD Peter Decherney
HOME Michael Allen Fox
HORMONES Martin Luck
HUMAN ANATOMY Leslie Klenerman
HUMAN EVOLUTION Bernard Wood
HUMAN RIGHTS Andrew Clapham
HUMANISM Stephen Law
HUME A. J. Ayer
HUMOUR Noël Carroll
THE ICE AGE Jamie Woodward
IDEOLOGY Michael Freeden
INDIAN CINEMA Ashish Rajadhyaksha
INDIAN PHILOSOPHY Sue Hamilton
THE INDUSTRIAL REVOLUTION Robert C. Allen
INFECTIOUS DISEASE Marta L. Wayne and Benjamin M. Bolker
INFINITY Ian Stewart
INFORMATION Luciano Floridi
INNOVATION Mark Dodgson and David Gann
INTELLIGENCE Ian J. Deary
INTELLECTUAL PROPERTY Siva Vaidhyanathan
INTERNATIONAL LAW Vaughan Lowe
INTERNATIONAL MIGRATION Khalid Koser
INTERNATIONAL RELATIONS Paul Wilkinson
INTERNATIONAL SECURITY Christopher S. Browning
IRAN Ali M. Ansari
ISLAM Malise Ruthven
ISLAMIC HISTORY Adam Silverstein
ISOTOPES Rob Ellam
ITALIAN LITERATURE Peter Hainsworth and David Robey
JESUS Richard Bauckham
JEWISH HISTORY David N. Myers
JOURNALISM Ian Hargreaves
JUDAISM Norman Solomon

JUNG Anthony Stevens
KABBALAH Joseph Dan
KAFKA Ritchie Robertson
KANT Roger Scruton
KEYNES Robert Skidelsky
KIERKEGAARD Patrick Gardiner
KNOWLEDGE Jennifer Nagel
THE KORAN Michael Cook
LANDSCAPE ARCHITECTURE Ian H. Thompson
LANDSCAPES AND GEOMORPHOLOGY Andrew Goudie and Heather Viles
LANGUAGES Stephen R. Anderson
LATE ANTIQUITY Gillian Clark
LAW Raymond Wacks
THE LAWS OF THERMODYNAMICS Peter Atkins
LEADERSHIP Keith Grint
LEARNING Mark Haselgrove
LEIBNIZ Maria Rosa Antognazza
LIBERALISM Michael Freeden
LIGHT Ian Walmsley
LINCOLN Allen C. Guelzo
LINGUISTICS Peter Matthews
LITERARY THEORY Jonathan Culler
LOCKE John Dunn
LOGIC Graham Priest
LOVE Ronald de Sousa
MACHIAVELLI Quentin Skinner
MADNESS Andrew Scull
MAGIC Owen Davies
MAGNA CARTA Nicholas Vincent
MAGNETISM Stephen Blundell
MALTHUS Donald Winch
MAMMALS T. S. Kemp
MANAGEMENT John Hendry
MAO Delia Davin
MARINE BIOLOGY Philip V. Mladenov
THE MARQUIS DE SADE John Phillips
MARTIN LUTHER Scott H. Hendrix
MARTYRDOM Jolyon Mitchell
MARX Peter Singer
MATERIALS Christopher Hall
MATHEMATICS Timothy Gowers
THE MEANING OF LIFE Terry Eagleton
MEASUREMENT David Hand
MEDICAL ETHICS Tony Hope
MEDICAL LAW Charles Foster
MEDIEVAL BRITAIN John Gillingham and Ralph A. Griffiths
MEDIEVAL LITERATURE Elaine Treharne
MEDIEVAL PHILOSOPHY John Marenbon
MEMORY Jonathan K. Foster
METAPHYSICS Stephen Mumford
THE MEXICAN REVOLUTION Alan Knight
MICHAEL FARADAY Frank A. J. L. James
MICROBIOLOGY Nicholas P. Money
MICROECONOMICS Avinash Dixit
MICROSCOPY Terence Allen
THE MIDDLE AGES Miri Rubin
MILITARY JUSTICE Eugene R. Fidell
MILITARY STRATEGY Antulio J. Echevarria II
MINERALS David Vaughan
MODERN ART David Cottington
MODERN CHINA Rana Mitter
MODERN DRAMA Kirsten E. Shepherd-Barr
MODERN FRANCE Vanessa R. Schwartz
MODERN IRELAND Senia Pašeta
MODERN ITALY Anna Cento Bull
MODERN JAPAN Christopher Goto-Jones
MODERN LATIN AMERICAN LITERATURE Roberto González Echevarría
MODERN WAR Richard English
MODERNISM Christopher Butler
MOLECULAR BIOLOGY Aysha Divan and Janice A. Royds
MOLECULES Philip Ball
THE MONGOLS Morris Rossabi
MOONS David A. Rothery
MORMONISM Richard Lyman Bushman
MOUNTAINS Martin F. Price
MUHAMMAD Jonathan A. C. Brown
MULTICULTURALISM Ali Rattansi
MULTILINGUALISM John C. Maher
MUSIC Nicholas Cook
MYTH Robert A. Segal
THE NAPOLEONIC WARS Mike Rapport

NATIONALISM Steven Grosby
NAVIGATION Jim Bennett
NELSON MANDELA Elleke Boehmer
NEOLIBERALISM Manfred Steger and Ravi Roy
NETWORKS Guido Caldarelli and Michele Catanzaro
THE NEW TESTAMENT Luke Timothy Johnson
THE NEW TESTAMENT AS LITERATURE Kyle Keefer
NEWTON Robert Iliffe
NIETZSCHE Michael Tanner
NINETEENTH-CENTURY BRITAIN Christopher Harvie and H. C. G. Matthew
THE NORMAN CONQUEST George Garnett
NORTH AMERICAN INDIANS Theda Perdue and Michael D. Green
NORTHERN IRELAND Marc Mulholland
NOTHING Frank Close
NUCLEAR PHYSICS Frank Close
NUCLEAR POWER Maxwell Irvine
NUCLEAR WEAPONS Joseph M. Siracusa
NUMBERS Peter M. Higgins
NUTRITION David A. Bender
OBJECTIVITY Stephen Gaukroger
OCEANS Dorrik Stow
THE OLD TESTAMENT Michael D. Coogan
THE ORCHESTRA D. Kern Holoman
ORGANIC CHEMISTRY Graham Patrick
ORGANIZATIONS Mary Jo Hatch
PAGANISM Owen Davies
PAIN Rob Boddice
THE PALESTINIAN-ISRAELI CONFLICT Martin Bunton
PANDEMICS Christian W. McMillen
PARTICLE PHYSICS Frank Close
PAUL E. P. Sanders
PEACE Oliver P. Richmond
PENTECOSTALISM William K. Kay
THE PERIODIC TABLE Eric R. Scerri
PHILOSOPHY Edward Craig
PHILOSOPHY IN THE ISLAMIC WORLD Peter Adamson
PHILOSOPHY OF LAW Raymond Wacks
PHILOSOPHY OF SCIENCE Samir Okasha
PHOTOGRAPHY Steve Edwards
PHYSICAL CHEMISTRY Peter Atkins
PILGRIMAGE Ian Reader
PLAGUE Paul Slack
PLANETS David A. Rothery
PLANTS Timothy Walker
PLATE TECTONICS Peter Molnar
PLATO Julia Annas
POLITICAL PHILOSOPHY David Miller
POLITICS Kenneth Minogue
POPULISM Cas Mudde and Cristóbal Rovira Kaltwasser
POSTCOLONIALISM Robert Young
POSTMODERNISM Christopher Butler
POSTSTRUCTURALISM Catherine Belsey
PREHISTORY Chris Gosden
PRESOCRATIC PHILOSOPHY Catherine Osborne
PRIVACY Raymond Wacks
PROBABILITY John Haigh
PROGRESSIVISM Walter Nugent
PROTESTANTISM Mark A. Noll
PSYCHIATRY Tom Burns
PSYCHOANALYSIS Daniel Pick
PSYCHOLOGY Gillian Butler and Freda McManus
PSYCHOTHERAPY Tom Burns and Eva Burns-Lundgren
PUBLIC ADMINISTRATION Stella Z. Theodoulou and Ravi K. Roy
PUBLIC HEALTH Virginia Berridge
PURITANISM Francis J. Bremer
THE QUAKERS Pink Dandelion
QUANTUM THEORY John Polkinghorne
RACISM Ali Rattansi
RADIOACTIVITY Claudio Tuniz
RASTAFARI Ennis B. Edmonds
THE REAGAN REVOLUTION Gil Troy
REALITY Jan Westerhoff
THE REFORMATION Peter Marshall
RELATIVITY Russell Stannard
RELIGION IN AMERICA Timothy Beal
THE RENAISSANCE Jerry Brotton

RENAISSANCE ART Geraldine A. Johnson
REVOLUTIONS Jack A. Goldstone
RHETORIC Richard Toye
RISK Baruch Fischhoff and John Kadvany
RITUAL Barry Stephenson
RIVERS Nick Middleton
ROBOTICS Alan Winfield
ROCKS Jan Zalasiewicz
ROMAN BRITAIN Peter Salway
THE ROMAN EMPIRE Christopher Kelly
THE ROMAN REPUBLIC David M. Gwynn
ROMANTICISM Michael Ferber
ROUSSEAU Robert Wokler
RUSSELL A. C. Grayling
RUSSIAN HISTORY Geoffrey Hosking
RUSSIAN LITERATURE Catriona Kelly
THE RUSSIAN REVOLUTION S. A. Smith
SAVANNAS Peter A. Furley
SCHIZOPHRENIA Chris Frith and Eve Johnstone
SCHOPENHAUER Christopher Janaway
SCIENCE AND RELIGION Thomas Dixon
SCIENCE FICTION David Seed
THE SCIENTIFIC REVOLUTION Lawrence M. Principe
SCOTLAND Rab Houston
SEXUALITY Véronique Mottier
SHAKESPEARE'S COMEDIES Bart van Es
SHAKESPEARE'S SONNETS AND POEMS Jonathan F. S. Post
SHAKESPEARE'S TRAGEDIES Stanley Wells
SIKHISM Eleanor Nesbitt
THE SILK ROAD James A. Millward
SLANG Jonathon Green
SLEEP Steven W. Lockley and Russell G. Foster
SOCIAL AND CULTURAL ANTHROPOLOGY John Monaghan and Peter Just
SOCIAL PSYCHOLOGY Richard J. Crisp
SOCIAL WORK Sally Holland and Jonathan Scourfield
SOCIALISM Michael Newman
SOCIOLINGUISTICS John Edwards
SOCIOLOGY Steve Bruce
SOCRATES C. C. W. Taylor
SOUND Mike Goldsmith
THE SOVIET UNION Stephen Lovell
THE SPANISH CIVIL WAR Helen Graham
SPANISH LITERATURE Jo Labanyi
SPINOZA Roger Scruton
SPIRITUALITY Philip Sheldrake
SPORT Mike Cronin
STARS Andrew King
STATISTICS David J. Hand
STEM CELLS Jonathan Slack
STRUCTURAL ENGINEERING David Blockley
STUART BRITAIN John Morrill
SUPERCONDUCTIVITY Stephen Blundell
SYMMETRY Ian Stewart
TAXATION Stephen Smith
TEETH Peter S. Ungar
TELESCOPES Geoff Cottrell
TERRORISM Charles Townshend
THEATRE Marvin Carlson
THEOLOGY David F. Ford
THINKING AND REASONING Jonathan St B. T. Evans
THOMAS AQUINAS Fergus Kerr
THOUGHT Tim Bayne
TIBETAN BUDDHISM Matthew T. Kapstein
TOCQUEVILLE Harvey C. Mansfield
TRAGEDY Adrian Poole
TRANSLATION Matthew Reynolds
THE TROJAN WAR Eric H. Cline
TRUST Katherine Hawley
THE TUDORS John Guy
TWENTIETH-CENTURY BRITAIN Kenneth O. Morgan
THE UNITED NATIONS Jussi M. Hanhimäki
THE U.S. CONGRESS Donald A. Ritchie
THE U.S. SUPREME COURT Linda Greenhouse
UTILITARIANISM Katarzyna de Lazari-Radek and Peter Singer
UTOPIANISM Lyman Tower Sargent
THE VIKINGS Julian Richards
VIRUSES Dorothy H. Crawford

VOLTAIRE Nicholas Cronk
WAR AND TECHNOLOGY
Alex Roland
WATER John Finney
WEATHER Storm Dunlop
THE WELFARE STATE David Garland
WILLIAM SHAKESPEARE
Stanley Wells
WITCHCRAFT Malcolm Gaskill
WITTGENSTEIN A. C. Grayling
WORK Stephen Fineman
WORLD MUSIC Philip Bohlman
THE WORLD TRADE
ORGANIZATION Amrita Narlikar
WORLD WAR II Gerhard L. Weinberg
WRITING AND SCRIPT
Andrew Robinson
ZIONISM Michael Stanislawski

Available soon:

PERCEPTION Brian J. Rogers
PROJECTS Andrew Davies
SOUTHEAST ASIA James R. Rush
BIG DATA Dawn E. Holmes
MIRACLES Yujin Nagasawa

For more information visit our website

www.oup.com/vsi/

Graham Priest

LOGIC

A Very Short Introduction

SECOND EDITION

Great Clarendon Street, Oxford, OX2 6DP,
United Kingdom

Oxford University Press is a department of the University of Oxford.
It furthers the University's objective of excellence in research, scholarship,
and education by publishing worldwide. Oxford is a registered trade mark of
Oxford University Press in the UK and in certain other countries

First edition published 2000
Second edition published 2017

Published in the United States of America by Oxford University Press
198 Madison Avenue, New York, NY 10016, United States of America

British Library Cataloguing in Publication Data
Data available

Library of Congress Control Number: 2017940856

ISBN 978-0-19-881170-1

Printed and bound by CPI Group (UK) Ltd, Croydon, CR0 4YY

This book is dedicated to all those who have ever thought about logic—or who ever will.

Contents

Preface to the first edition

Logic is one of the most ancient intellectual disciplines, and one of the most modern. Its beginnings go back to the 4th century BCE. The only older disciplines are philosophy and mathematics, with both of which it has always been intimately connected. It was revolutionized around the turn of the 20th century, by the application of new mathematical techniques, and in the last half-century it has found radically new and important roles in computation and information processing. It is thus a subject that is central to much human thought and endeavour.

This book is an introduction to logic, as contemporary logicians now understand the subject. It does not attempt to be a textbook, however. There are numerous such books currently available. The point of this one is to explore the roots of logic, which sink deep into philosophy. Some formal logic will be explained along the way.

In each of the main chapters, I start by taking up some particular philosophical problem or logical puzzle. I then explain one approach to it. Often this is a fairly standard one; but in some of the areas there is no standard answer: logicians still disagree. In such cases, I have just chosen one that is interesting. Nearly all the approaches, whether standard or not, may be challenged. I finish each chapter with some problems for the approach that I have explained. Sometimes these problems are standard; sometimes they are not. Sometimes they may have easy answers; sometimes they may not. The aim is to challenge you to figure out what you make of the matter.

Modern logic is a highly mathematical subject. I have tried to write the material in such a way as to avoid nearly all mathematics. The most that is required is a little high-school algebra in Chapters 11–13. It is true that you will need the determination to master some symbolism that may be new to you; but this is a lot less than is required to have a basic grasp of any new language. And the perspicuity that the symbolism gives to difficult questions makes any trouble one may have in mastering it well worth it. One warning, though: reading a book on logic or philosophy is not like reading a novel. There will be times when you will have to read slowly and carefully. Sometimes you may have to stop and think about things; and you should be prepared to go back and reread a paragraph if necessary.

The appendix 'A little history and some further reading' at the end of the book is on the development of logic. In this, I have tried to put some of the issues that the book deals within a historical perspective, to show that logic is a living subject, which has always evolved, and which will continue to do so. The section also contains suggestions for further reading.

There are three other appendices. The first contains a glossary of terms and symbols. You may consult this if you forget the meaning of a word or symbol. The second contains a question relevant to each chapter, with which you can test your understanding of its main ideas. The third contains solutions to the problems. This appendix was not in the early printings of the first edition, but was added to later printings.

The book goes for breadth rather than depth. It would be easy to write a book on the topic of every single chapter—indeed, many such books have been written. And even so, there are very many important issues in logic that I have not even touched on here. But if you hang in there till the end of the book, you will have a pretty good idea of the fundamentals of modern logic, and why people find it worth thinking about the subject.

Preface to the second edition

The first edition of this book appeared in 2000. In 2016, Oxford University Press approached me concerning the possibility of a second edition. My initial reaction was that there was not a lot of point in this. If this had been a book on, say, international relations, it would certainly have been out of date by now. But logic develops, generally speaking, at a more measured pace. The material in the first edition is as good now as it was then. And I don't think I can do a much better job of covering it for the intended audience. Finally, the book has been selling well, now having been translated into eight other languages (if I have kept count correctly).

However, the Press pointed out to me that I had not used up all of my allotted 35,000 words in the first edition, and suggested that I might add a couple of new chapters. After pondering what I might add, this struck me as a very good idea. The first edition gives a sense of the basics of modern logic; but no more. Logic is a subject of great depth and beauty, and the material in the first edition gives no real sense of this. Two extra chapters would allow me to do so—at least to a limited extent: there is no hope of doing more than scratch the surface of the matter in a book of this nature. However, the new chapters do give the reader a 'glimpse beyond'—or so I hope.

So the present book contains the original chapters, pretty much as they were, with some occasional smoothing. The historical Chapter 14 has now become 'A little history and some further reading'; and there are two new chapters—14 and 15—which deal, respectively, with Turing and the Halting Theorem, and Gödel and his Incompleteness Theorems. I have maintained the format of the first edition. The key points of each chapter are itemized at its end. The historical material, problems, and glossary, have been extended to cover the new material.

The new chapters are perforce harder than previous ones; but I have done my best to make them as user-friendly as possible. Those who omit these chapters will be as well off as those who purchased the first edition; and those who venture into these two chapters will, I hope, get at least some sense of where the more elementary material can lead.

List of illustrations

Chapter 1
Validity: what follows from what?

Most people like to think of themselves as logical. Telling someone 'You are not being logical' is normally a form of criticism. To be illogical is to be confused, muddled, irrational. But what is logic? In Lewis Carroll's *Through the Looking Glass*, Alice meets the logic-chopping pair Tweedledum and Tweedledee (Figure 1). When Alice is lost for words, they go onto the attack:

> 'I know what you are thinking about', said Tweedledum: 'but it isn't so, nohow.'
>
> 'Contrariwise,' continued Tweedledee, 'if it was so, it might be; and if it were so, it would be: but as it isn't, it ain't. That's logic.'

What Tweedledee is doing—at least, in Carroll's parody—is reasoning. And that, as he says, is what logic is about.

We all reason. We try to figure out what is so, reasoning on the basis of what we already know. We try to persuade others that something is so by giving them reasons. Logic is the study of what counts as a good reason for what, and why. You have to

1. Tweedledum and Tweedledee debate the finer points of logic with Alice.

understand this claim in a certain way, though. Here are two bits of reasoning—logicians call them *inferences*:

1. Rome is the capital of Italy, and this plane lands in Rome; so the plane lands in Italy.
2. Moscow is the capital of the USA; so you can't go to Moscow without going to the USA.

In each case, the claims before the 'so'—logicians call them *premisses*—are giving reasons; the claims after the 'so'—logicians call them *conclusions*—are what the reasons are supposed to be reasons for. The first piece of reasoning is fine; but the second is pretty hopeless, and wouldn't persuade anyone with an elementary knowledge of geography: the premiss, that Moscow is the capital of the USA, is simply false. Notice, though, that if the premiss had been true—if, say, the USA had bought the whole of Russia (not just Alaska) and had moved the White House to Moscow to be nearer the centres of power in Europe—the conclusion would indeed have been true. It would have followed from the premisses; and that is what logic is concerned with. It is not concerned with whether the premisses of an inference are true or false. That's somebody else's business (in this case, the geographer's). It is interested simply in whether the conclusion follows from the premisses. Logicians call an inference where the conclusion really does follow from the premisses *valid*. So the central aim of logic is to understand validity.

You might think this a rather dull task—an intellectual exercise with somewhat less appeal than solving crossword puzzles. But it turns out that this is not only a very hard matter; it is one that cannot be divorced from a number of important (and sometimes profound) philosophical questions. We will see some of these as we go along. For the moment, let us get a few more of the basic facts about validity straight.

To start with, it is common to distinguish between two different kinds of validity. To understand this, consider the following three inferences:

1. If the burglar had broken in through the kitchen window, there would be footprints outside; but there are no footprints; so the burglar didn't break in through the kitchen window.
2. Storm clouds are gathering; so there will be rain.
3. Storm clouds are gathering; so the burglar didn't break in through the kitchen window.

The first inference is a very straightforward one. If the premisses are true, so must the conclusion be. Or, to put it another way, the premisses couldn't be true without the conclusion also being true. Logicians call an inference of this kind *deductively valid*. Inference number two is a bit different. The premiss clearly gives a good reason for the conclusion, but it is not completely conclusive. After all, a change of wind can sometimes take the clouds in a different direction. So the inference is not deductively valid. Inferences like this are usually said to be *inductively valid.* Inference number three, by contrast, appears pretty hopeless by any standard. The premiss seems to provide no kind of reason for the conclusion at all. It is invalid—both deductively and inductively. In fact, since people are not complete idiots, if someone actually offered a reason like this, one would assume that there is some extra premiss that they had not bothered to tell us (maybe that burglars don't like to risk standing in the rain getting wet).

Inductive validity is a very important notion. We reason inductively all the time; for example, in trying to solve problems such as why the car has broken down, why a person is ill, or who committed a crime. The fictional logician Sherlock Holmes was a master of it. Despite this, historically, much more effort has gone into understanding deductive validity—maybe because logicians have tended to be philosophers or mathematicians (in whose studies deductively valid inferences are centrally important), and not doctors or detectives. We will come back to the notion of induction later in the book. For the present, let's think some more about deductive validity. (It is natural to suppose that deductive validity is the simpler notion, since valid inferences are more cut-and-dried. So it's not a bad idea to try to understand this first. That, as we shall see, is hard enough.) Until further notice 'valid' will simply mean 'deductively valid'.

So what is a valid inference? One, we saw, where the premisses can't be true without the conclusion also being true. But what does that mean? In particular, what does the *can't* mean? In general, 'can't' can mean many different things. Consider, for example: 'Mary can play the piano, but John can't'; here we are talking about human abilities. Compare: 'You can't go in here: you need a permit'; here we are talking about what some code of rules permits.

It is natural to understand the 'can't' relevant to the present case in this way: to say that the premisses can't be true without the conclusion being true is to say that in all situations in which all the premisses are true, so is the conclusion. So far so good; but what, exactly, is a situation? What sorts of things go into their makeup, and how do these things relate to each other? And what is it to be *true*? Now, there's a philosophical problem for you, as Tweedledee might have said.

These issues will concern us by and by; but let us leave them for the time being, and finish with one more thing. One shouldn't run away with the idea that the explanation of deductive validity that I have just given is itself unproblematic. (In philosophy, all interesting claims are contentious.) Here is one problem. Assuming that the account is correct, to know that an inference is deductively valid is to know that there are no situations in which the premisses are true and the conclusion is not. Now, on any reasonable understanding of what it is to be a situation, there are an awful lot of them: situations about things on the planets of distant stars; situations about events before there were any living beings in the cosmos; situations described in works of fiction; situations imagined by visionaries. How can one know what holds in *all* situations? Worse, there would appear to be an infinite number of situations (situations one year hence, situations two years hence, situations three years hence,...). It is therefore impossible, even in principle, to survey all situations. So if this account of validity is correct, and given that we can recognize inferences as valid or invalid (at least in many cases) we must have some insight into this, from some special source. What source?

Do we need to invoke some sort of mystic intuition? Not necessarily. Consider an analogous problem. We can all distinguish between grammatical and ungrammatical strings of words of our native language without too much problem. For example, any native speaker of English would recognize that

'This is a chair' is a grammatical sentence while 'A chair is is a' is not. But there would appear to be an infinite number of both grammatical and ungrammatical sentences. (For example, 'One is a number', 'Two is a number', 'Three is a number',... are all grammatical sentences. And it is easy enough to produce word salads *ad libitum*). So how do we do it? Perhaps the most influential of modern linguists, Noam Chomsky, suggested that we can do this because the infinite collections are encapsulated in a finite set of rules that are hard-wired into us; that evolution has programmed us with an innate grammar. Could logic be the same? Are the rules of logic hard-wired into us in the same way?

Main ideas of the chapter

- A valid inference is one where the conclusion follows from the premiss(es).
- A deductively valid inference is one for which there is no situation in which all the premisses are true, but the conclusion is not.

Chapter 2
Truth functions—or not?

Whether or not the rules of validity are hard-wired into us, we all have pretty strong intuitions about the validity or otherwise of various inferences. There wouldn't be much disagreement, for example, that the following inference is valid: 'She's a woman and a banker; so she's a banker'. Or that the following inference is invalid: 'He's a carpenter; so he's a carpenter and plays baseball'.

But our intuitions can get us into trouble sometimes. What do you think of the following inference? The two premisses occur above the line; the conclusion below it.

The Queen is rich.   The Queen isn't rich.

Pigs can fly.

It certainly doesn't seem valid. The wealth of the Queen—great or not—would seem to have no bearing on the aviatory abilities of pigs.

But what do you think about the following two inferences?

The Queen is rich.

Either the Queen is rich or pigs can fly.

Either the Queen is rich or pigs can fly.   The Queen isn't rich.

Pigs can fly.

The first of these seems valid. Consider its conclusion. Logicians call sentences like this a *disjunction*; and the clauses on either side of the 'or' are called *disjuncts*. Now, what does it take for a disjunction to be true? Just that one or other of the disjuncts is true. So in any situation where the premiss is true, so is the conclusion. The second inference also seems valid. If one or other of two claims is true and one of these isn't, the other must be.

Now, the trouble is that by putting these two apparently valid inferences together, we get the apparently invalid inference, like this:

$$\dfrac{\dfrac{\text{The Queen is rich.}}{\text{Either the Queen is rich or pigs can fly.}} \quad \text{The Queen isn't rich.}}{\text{Pigs can fly.}}$$

This can't be right. Chaining valid inferences together in this way can't give you an invalid inference. If all the premisses are true in any situation, then so are their conclusions, the conclusions that follow from *these*; and so on, till we reach the final conclusion. What has gone wrong?

To give an orthodox answer to this question, let us focus a bit more on the details. For a start, let's write the sentence 'Pigs can fly' as p, and the sentence 'The Queen is rich' as q. This makes things a bit more compact; but not only that: if you think about it for a moment, you can see that the two particular sentences actually used in the examples above don't have much to do with anything; I could have set everything up using pretty much any two sentences; so we can ignore their content. This is what we do in writing the sentences as single letters.

The sentence 'Either the Queen is rich or pigs can fly' now becomes 'Either q or p'. Logicians often write this as $q \vee p$. What of 'The Queen isn't rich'? Let us rewrite this as 'It is not the case that the Queen is rich', pulling the negative particle to the front of the sentence. Hence, the sentence becomes 'It is not the case that *q*'. Logicians often write this as $\neg q$, and call it the *negation* of *q*. While we are at it, what about the sentence 'The Queen is rich *and* pigs can fly', that is, '*q* and *p*'? Logicians often write this as $q \& p$ and call it the *conjunction* of *q* and *p*, *q* and *p* being the *conjuncts*. With this machinery under our belt, we can write the chain-inference that we met thus:

$$\dfrac{\dfrac{q}{q \vee p} \quad \neg q}{p}$$

What are we to say about this inference?

Sentences can be true, and sentences can be false. Let us use *T* for truth, and *F* for falsity. Following one of the founders of modern logic, the German philosopher/mathematician Gottlob Frege (Figure 2), these are often called *truth values*. Given any old sentence, *a*, what is the connection between the truth value of *a* and that of its negation, $\neg a$? A natural answer is that if one is true, the other is false, and vice versa. Thus, if 'The Queen is rich' is true, 'The Queen isn't rich' is false, and vice versa. We can record this as follows:

> $\neg a$ has the value *T* just if *a* has the value *F*.
> $\neg a$ has the value *F* just if *a* has the value *T*.

Logicians call these the *truth conditions* for negation. If we assume that every sentence is either true or false, but not both, we can depict the conditions in the following table, which logicians call a *truth table*:

a	$\neg a$
T	F
F	T

2. Gottlob Frege (1848–1925), one of the founders of modern logic.

If a has the truth value given in the column under it, $\neg a$ has the corresponding value to its right.

What of disjunction, $\vee$? As I have already noted, a natural assumption is that a disjunction, $a \vee b$, is true if one or other (or maybe both) of a and b are true, and false otherwise. We can record this in the truth conditions for disjunction:

$a \vee b$ has the value T just if at least one of a and b has the value T.
$a \vee b$ has the value F just if both of a and b have the value F.

These conditions can be depicted in the following truth table:

a	b	$a \vee b$
T	T	T
T	F	T
F	T	T
F	F	F

Each row—except the first, which is the header—now records a possible combination of the values for a (first column) and b (second column). There are four such possible combinations, and so four rows. For each combination, the corresponding value of $a \vee b$ is given to its right (third column).

Again, while we are about it, what is the connection between the truth values of a and b, and that of $a \& b$? A natural assumption is that $a \& b$ is true if both a and b are true, and false otherwise. Thus, for example, 'John is 35 and has brown hair' is true just if 'John is 35' and 'John has brown hair' are both true. We can record this in the truth conditions for conjunction:

> $a \& b$ has the value T just if both of a and b have the value T.
> $a \& b$ has the value F just if at least one of a and b has the value F.

These conditions can be depicted in the following truth table:

a	b	$a \& b$
T	T	T
T	F	F
F	T	F
F	F	F

Now, how does all this bear on the problem we started with? Let us come back to the question I raised towards the end of Chapter 1: what is a situation? A natural thought is that whatever a situation is, it determines a truth value for every sentence. So, for example, in one particular situation, it might be true that the Queen is rich and false that pigs can fly. In another it might be false that the Queen is rich, and true that pigs can fly. (Note that these situations may be purely hypothetical!) In other words, a situation determines each relevant sentence to be either *T* or *F*. The relevant sentences here do not contain any occurrences of 'and', 'or', or 'not'. Given the basic information about a situation, we can use truth tables to work out the truth values of the sentences that do.

For example, suppose we have the following situation:

$p : T$
$q : F$
$r : T$

(r might be the sentence 'Rhubarb is nutritious', and '$p : T$' means that p is assigned the truth value T, etc.) What is the truth value of, say, $p \& (\neg r \vee q)$? We work out the truth value of this in exactly the same way that we would work out the numerical value of $3 \times (-6 + 2)$ using tables for multiplication and addition. The truth value of r is T. So the truth table for $\neg$ tells us that the truth value of $\neg r$ is F. But since the value of q is F, the truth table for $\vee$ tells us that the value of $\neg r \vee q$ is F. And since the truth value of p is T, the truth table for & tells us that the value of $p \& (\neg r \vee q)$ is F. In this step-by-step way, we can work out the truth value of any formula containing occurrences of &, $\vee$, and $\neg$.

Now, recall from Chapter 1 that an inference is valid provided that there is no situation which makes all the premisses true, and the conclusion untrue (false). That is, it is valid if there is no way of assigning *T*s and *F*s to the relevant sentences, which results in all the premisses having the value T and the conclusion having the value F. Consider, for example, the inference that we have already met, $q \,/\, q \vee p$. (I write this on a single line to save Oxford University Press money.) The relevant sentences are q and p. There are four combinations of truth values, and for each of these we can work out the truth values for the premiss and conclusion. We can represent the result as follows:

q	p	q	$q \vee p$
T	T	T	T
T	F	T	T
F	T	F	T
F	F	F	F

The first two columns give us all the possible combinations of truth values for q and p. The last two columns give us the corresponding truth values for the premiss and the conclusion. The third column is the same as the first. This is an accident of this example, due to the fact that, in this particular case, the premiss happens to be one of the relevant sentences. The fourth column can be read off from the truth table for disjunction. Given this information, we can see that the inference is valid. For there is no row where the premiss, q, is true and the conclusion, $q \vee p$, is not.

What about the inference $q \vee p, \neg q \,/\, p$? Proceeding in the same way, we obtain:

q	p	$q \vee p$	$\neg q$	p
T	T	T	F	T
T	F	T	F	F
F	T	T	T	T
F	F	F	T	F

This time, there are five columns, because there are two premisses. The truth values of the premisses and conclusion can be read off from the truth tables for disjunction and negation. And again, there is no row where both of the premisses are true and the conclusion is not. Hence, the inference is valid.

What about the inference with which we started: $q, \neg q / p$? Proceeding as before, we get:

q	p	q	$\neg q$	p
T	T	T	F	T
T	F	T	F	F
F	T	F	T	T
F	F	F	T	F

Again, the inference is valid; and now we see why. There is no row in which both of the premisses are true and the conclusion is false. Indeed, there is no row in which both of the premisses are true. The conclusion doesn't really matter at all! Sometimes, logicians describe this situation by saying that the inference is *vacuously* valid, just because the premisses could never be true together.

Here, then, is a solution to the problem with which we started. According to this account, our original intuitions about this inference were wrong. After all, people's intuitions can often be misleading. It seems obvious to everyone that the Earth is motionless—until they take a course in physics, and find out that it is really hurtling through space. We can even offer an explanation as to why our logical intuitions go wrong. Most of the inferences we meet in practice are not of the vacuous kind. Our intuitions develop in this sort of context, and don't apply generally—just as the habits you build up learning to walk (for example, not to lean to the side) don't always work in other contexts (for example, when you to learn to ride a bike).

We will come back to this matter in a later chapter. But let us end this one with a brief look at the adequacy of the machinery we have used. Things here are not as straightforward as one might have hoped. According to this account, the truth value of a sentence $\neg a$ is completely determined by the truth value of the sentence a. In a similar way, the truth values of the sentences $a \vee b$ and $a \& b$ are completely determined by the truth values of a and b. Logicians call operations that work like this *truth functions*. But there are good reasons to suppose that 'or' and 'and', as they occur in English, are not truth functions—at least, not always.

For example, according to the truth table for &, '*a* and *b*' always has the same truth value as '*b* and *a*': namely, they are both true if *a* and *b* are both true, and false otherwise. But consider the sentences:

1. John hit his head and fell down.
2. John fell down and hit his head.

The first says that John hit his head *and then* fell down. The second says that John fell down *and then* hit his head. Clearly, the first could be true whilst the second is false, and vice versa. Thus, it is not just the truth values of the conjuncts that are important, but which conjunct caused which.

Similar problems beset 'or'. According to the account we had, '*a* or *b*' is true if one or other of *a* and *b* is true. But suppose a friend says:

> Either you come now or we will be late;

and so you come. Given the truth table for $\vee$, the disjunction is true. But suppose you discover that your friend had been tricking you: you could have left half an hour later and still been on time. Under these circumstances, you would surely say that your friend had lied: what he had said was false. Again, it is not merely the truth values of the disjuncts that are important, but the existence of a connection of a certain kind between them.

I will leave you to think about these matters. The material we have been looking at gives us at least a working account of how certain logical machinery functions; and we will draw on this in succeeding chapters, unless the ideas in those chapters explicitly override it—which they will sometimes.

The machinery in question deals only with certain kinds of inferences: there are many others. We have only just started.

Main ideas of the chapter

- In a situation, a unique truth value (T or F) is assigned to each relevant sentence.
- $\neg a$ is T just if a is F.
- $a \vee b$ is T just if at least one of a and b is T.
- $a \& b$ is T just if both of a and b are T.

Chapter 3

Names and quantifiers: is nothing something?

The inferences that we looked at in Chapter 2 involved phrases like 'or' and 'it is not the case that', words that add to, or join, whole sentences to make other whole sentences; but there are lots of inferences that appear to work in a quite different way. Consider, for example, the inference:

Marcus gave me a book.
―――――――――――――――
Someone gave me a book.

Neither the premiss nor the conclusion has a part which is itself a whole sentence. If this inference is valid, it is so because of what is going on *within* whole sentences.

Traditional grammar tells us that the simplest whole sentences are composed of a *subject* and a *predicate*. Thus, consider the examples:

1. Marcus saw the elephant.
2. Annika fell asleep.
3. Someone hit me.
4. Nobody came to my party.

The first word, in each case, is the subject of the sentence: it tells us what the sentence is about. The rest is the predicate: this tells us what is said about it. Now, when is such a sentence true?

Take the second example. It is true if the object referred to by the subject 'Annika' has the property expressed by the predicate, that is, fell asleep.

All well and good. But what does the subject of sentence 3 refer to? The person who hit me? But maybe nobody hit me. No one said that this was a true sentence. The case with sentence 4 is even worse. To whom does 'nobody' refer? In *Through the Looking Glass*, just before her encounter with the Lion and the Unicorn, Alice comes across the White King, who is waiting for a messenger. (For some reason, when the messenger turns up, it looks disconcertingly like a rabbit (Figure 3).) When the King meets Alice, he says:

> 'Just look along the road, and tell me if you can see... [the Messenger].'
>
> 'I see nobody on the road,' said Alice.
>
> 'I only wish I had such eyes,' the King remarked in a fretful tone. 'To be able to see Nobody! And at that distance too! Why, it's as much as *I* can do to see real people, by this light!'

Carroll is making a logical joke, as he often does. When Alice says that she can see nobody, she is not saying that she can see a person—real or otherwise. 'Nobody' does not refer to a person—or to anything else.

Words like 'nobody', 'someone', 'everyone' are called by modern logicians *quantifiers*, and they are distinguished from names like 'Marcus' and 'Annika'. What we have just seen is that, even if both quantifiers and names can be the grammatical subjects of sentences, they must function in quite different ways. So, how do quantifiers work?

Here is a standard modern answer. A situation comes furnished with a stock of objects. In our case, the relevant objects are all people.

3. Nobody.

All the names which occur in our reasoning about this situation refer to one of the objects in this collection. Thus, if we write m for 'Marcus', m refers to one of these objects. And if we write H for 'is happy', then the sentence mH is true in the situation just if the object referred to by m has the property expressed by H. (For perverse reasons of their own, logicians usually reverse the order, and write Hm, instead of mH. This is just a matter of convention.)

Now consider the sentence 'Someone is happy'. This is true in the situation just if there is *some* object or other, in the collection of objects, that is happy—that is, some object in the collection, call it x, is such that x is happy. Let us write 'Some object, x, is such that' as $\exists x$. Then we may write the sentence as: '$\exists x\ x$ is happy'; or remembering that we are writing 'is happy' as H, as: $\exists x\, xH$. Logicians sometimes call $\exists x$ a *particular quantifier*.

What about 'Everyone is happy'? This is true in a situation if *every* object in the relevant collection is happy. That is, every object, x, in the collection is such that x is happy. If we write 'Every object, x, is such that' as $\forall x$, then we can write this as $\forall x\, xH$. Logicians usually call $\forall x$ a *universal quantifier.*

There are now no prizes for guessing how we are to understand 'Nobody is happy'. This just means that there is no object, x, in the relevant collection, such that x is happy. We could have a special symbol meaning 'No object, x, is such that', but as a matter of fact, logicians don't normally bother with one. For to say that no one is happy is to say it is not the case that somebody is happy. So we may write this as $\neg\exists x\, xH$.

This analysis of quantifiers shows us that names and quantifiers work quite differently. In particular, the fact that 'Marcus is happy' and 'Someone is happy' get written, very differently, as mH and $\exists x\, xH$, respectively, shows us this. It shows us, moreover, that apparently simple grammatical form may be misleading. Not all grammatical subjects are equal. The account, incidentally, shows us why the inference with which we started is valid. Let us write G for 'gave me the book'. Then the inference is:

$$\frac{mG}{\exists x\, xG}$$

It is clear that if, in some situation, the object referred to by the name m gave me the book, then some object in the relevant collection gave me the book. By contrast, the White King is inferring from the fact that Alice saw nobody that she saw somebody (viz., Nobody). If we write 'is seen by Alice' as A then the King's inference is:

$$\frac{\neg\exists x\, xA}{\exists x\, xA}$$

This is clearly invalid. If there is no object in the relevant domain that was seen by Alice, it is obviously *not* true that there is some object in the relevant domain that was seen by her.

You might think that this is all a lot of fuss about nothing—in fact, just a way of spoiling a good joke. But it's a lot more serious than that. For quantifiers play a central role in many important arguments in mathematics and philosophy. Here is one philosophical example. It's a natural assumption that nothing happens without an explanation: people don't get ill for no reason; cars don't break down without a fault. Everything, then, has a cause. But what could the cause of everything be? Obviously it can't be anything physical, like a person; or even something like the Big Bang of cosmology. Such things must themselves have causes. So it must be something metaphysical. God is the obvious candidate.

This is one version of an argument for the existence of God, often called the *Cosmological Argument*. One might object to the argument in various ways. But at its heart, there is an enormous logical fallacy. The sentence 'Everything has a cause' is ambiguous. It can mean that everything that happens has some cause or other—that is, for every x, there is a y, such that x is caused by y; or it can mean that there is something which is the cause of everything—that is, there is some y such that for every x, x is caused by y. Suppose we think of the relevant domain of objects as causes and effects, and write 'x is caused by y' as xCy. Then we can write these two meanings as, respectively:

1 $\forall x\ \exists y\ xCy$

2 $\exists y\ \forall x\ xCy$

Now, these are not logically equivalent. The first follows from the second. If there is a thing which is the cause of *everything*, then certainly, everything that happens has *some cause or other*. But if everything has some cause or other, it does not follow that there is one and the same thing which is the cause of everything. (Compare: Everyone has a mother; it does not follow that there is someone who is the mother of everyone.)

This version of the Cosmological Argument trades on this ambiguity. What is established by talk of illnesses and cars is 1. But immediately, the argument goes on to ask what that cause is, assuming that it is 2 that has been established. Moreover, this slide is hidden because, in English, 'Everything has a cause' can be used to express either 1 or 2. Notice, also, that there is no ambiguity if the quantifiers are replaced by names. 'The background radiation of the cosmos is caused by the Big Bang' is not at all ambiguous. It may well be that a failure to distinguish between names and quantifiers is a further reason why one may fail to see the ambiguity.

So a correct understanding of quantifiers is important—and not just for logic. The words 'something', 'nothing', etc., do not stand for objects, but function in a completely different way. Or at least, they can do: things are not quite that simple. Consider the cosmos again. Either it stretches back infinitely into time past, or at some particular time it came into existence. In the first case, it had no beginning, but was always there; in the second, it began at some particular time. At different times, physics has, in fact, told us different things about the truth of this matter. Never mind this, however; just consider the second possibility. In this case, the cosmos came into existence out of nothing—or nothing physical, anyway, the cosmos being the totality of everything physical. Now consider that sentence, 'The cosmos came into existence out of nothing'. Let c be the cosmos, and let us write 'x came into existence out of y' as xEy. Then given our understanding of quantifiers, this sentence should mean $\neg\exists x\, cEx$. But it does not mean this; for this is equally true in the first alternative cosmology. In this, the cosmos, being infinite in time past, did not come into being at all. In particular, then, it is not the case that it came into being *from* something or other. When we say that in the second cosmology the cosmos came into existence out of nothing, we mean that it came into being from *nothingness*. So nothing *can* be a thing. Perhaps the White King was not so foolish after all.

Main ideas of the chapter

- The sentence nP is true in a situation if the object referred to by n has the property expressed by P in that situation.
- $\exists x\, xP$ is true in a situation just if some object in the situation, x, is such that xP.
- $\forall x\, xP$ is true in a situation just if every object in the situation, x, is such that xP.

Chapter 4
Descriptions and existence: did the Greeks worship Zeus?

While we are on the topic of subjects and predicates, there is a certain kind of phrase that can be the subject of sentences, which we haven't talked about yet. Logicians usually call them *definite descriptions*, or sometimes just *descriptions*—though be warned that this is a technical term. Descriptions are phrases like 'the man who first landed on the Moon' and 'the only man-made object on the Earth that is visible from space'. In general, descriptions have the form: *the thing satisfying such and such a condition*. Following the English philosopher/mathematician, Bertrand Russell (Figure 4), one of the founders of modern logic, we can write them as follows. Rewrite 'the man who first landed on the Moon' as 'the object, x, such that x is a man and x landed first on the Moon'. Now write ιx for 'the object, x, such that', and this becomes 'ιx(x is a man and x landed first on the Moon)'. If we write M for 'is a man' and F for 'landed first on the Moon', we then get: $\iota x(xM \,\&\, xF)$. In general, a description is something of the form $\iota x c_x$, where c_x is some condition containing occurrences of x. (That's what the little subscript x is there to remind you of.)

Since descriptions are subjects, they can be combined with predicates to make whole sentences. Thus, if we write U for 'was born in the USA', then 'the man who first landed on the Moon was born in the USA' is: $\iota x(xM \,\&\, xF)U$. Let us write μ as a shorthand for $\iota x(xM \,\&\, xF)$. (I use a Greek letter to remind you

4. Bertrand Russell (1872–1970), another of the founders of modern logic.

that it is really a description.) Then this is μU. Similarly, 'The first man to land on the Moon is a man and he landed first on the Moon' is μM & μF.

In terms of the division of Chapter 3, descriptions are names, not quantifiers. That is, they refer to objects—if we are lucky: we'll come back to that. Thus, 'The man who first landed on the Moon was born in the USA', μU, is true just if the particular person referred to by the phrase μ has the property expressed by U.

But descriptions are a special kind of name. Unlike what we might call *proper names*, like 'Annika' and 'the Big Bang', they carry information about the object referred to. Thus, for example, 'the man who first landed on the Moon' carries the information that the object referred to has the property of being a man and being first on the Moon. This might all seem banal and obvious, but things are not as simple as they appear. Because descriptions carry information in this way, they are often central to important arguments in mathematics and philosophy; and one way to appreciate some of these complexities is to look at an example of such an argument. This is another argument for the existence of God, often called the *Ontological Argument*. The argument comes in a number of versions, but here is a simple form of it:

> God is the being with all the perfections.
> But existence is a perfection.
> So God possesses existence.

In other words, God exists. If you haven't met this argument before, it will appear rather puzzling. For a start, what is a perfection? Loosely, a perfection is something like omniscience (knowing everything that there is to know), omnipotence (being able to do everything that can be done), and being morally perfect (acting always in the best possible way). In general, the perfections are all those properties that it is a jolly good thing to have. Now, the second premiss says that existence is a perfection. Why on earth should this be so? The reason one might suppose this to be so is a rather complex one, with its roots in the philosophy of one of the two most influential ancient Greek philosophers, Plato. Fortunately, we can work around this issue. We can make a list of properties like omniscience, omnipotence, etc., include existence in the list, and simply let 'perfection' mean any property on the list. Moreover, we can take 'God' to be synonymous with a certain description, namely, 'the being which has all the perfections (i.e. those properties on the list)'. In the Ontological Argument, both premisses are now true by definition and so drop out of the picture. The Argument then reduces to a one-liner:

> The object which is omniscient, omnipotent, morally perfect... and exists, exists

—and, we might add, is omnipotent, omniscient, morally perfect, and so on. This certainly looks to be true. To make things more perspicuous, suppose we write the list of God's properties as $P_1, P_2, \ldots, P_n$. So the last one, P_n, is existence. The definition of 'God' is: $\iota x(xP_1 \& \ldots \& xP_n)$. Let us write this as γ. Then the one-liner is $\gamma P_1 \& \ldots \& \gamma P_n$ (from which γP_n follows).

This is a special case of something more general, namely: *the thing satisfying such and such a condition, satisfies that very condition.* This is often called the *Characterization Principle* (a thing has those properties by which it is characterized). We'll abbreviate this as CP. We have already met an example of the CP, with 'The first man to land on the Moon is a man and he landed first on the moon', $\mu M \& \mu F$. In general, we obtain a case of the CP if we take some description, $\iota x c_x$, and substitute it for every occurrence of x in the condition c_x.

Now, for all the world, the CP looks to be true by definition. Of course things have those properties that they are characterized as having. Unfortunately, in general, it is false. For many things follow from it that are indisputably untrue.

For a start, we can use it to deduce the existence of all kinds of things that do not really exist. Consider the (non-negative) integers: 0, 1, 2, 3,.... There is no greatest. But using the CP, we can show the existence of a greatest. Let c_x be the condition 'x is the greatest integer & x exists'. Let δ be $\iota x c_x$. Then the CP gives us 'δ is the greatest integer, and δ exists'. The absurdities do not end there. Consider some unmarried person, say the Pope. We can prove that he is married. Let c_x be the condition 'x married the Pope'. Let δ be the description $\iota x c_x$. The CP gives us 'δ married the Pope'. So someone married the Pope, in other words the Pope is married.

What is to be said about all this? A fairly standard modern answer goes as follows. Consider the description $\iota x c_x$. If there is a unique object that satisfies the condition c_x in some situation, then the description refers to it. Otherwise, it refers to nothing: it is an 'empty name'. Thus, there is a unique x, such that x is a man and x landed first on the moon: Armstrong. So 'the x such that x is a man and x landed first on the moon' refers to Armstrong. Similarly, there is a unique least integer, namely 0; hence, the description 'the object which is the least integer' denotes 0. But since there is no greatest integer, 'the object which is the greatest integer' fails to refer to anything. Similarly, the description 'the city in Australia which has more than a million people' also fails to refer. Not, this time, because there are no such cities, but because there are several of them.

What has this to do with the CP? Well, if there is a unique object satisfying c_x in some situation, then $\iota x c_x$ refers to it. So the instance of the CP concerning c_x is true: $\iota x c_x$ is one of the things—in fact, the only thing—that satisfies c_x. In particular, the least integer is (indeed) the least integer; the city which is the federal capital of Australia, is indeed, the federal capital of Australia, etc. So some instances of the CP hold.

But what if there is no unique object satisfying c_x? If n is a name and P is a predicate, the sentence nP is true just if there is an object that n refers to, and it has the property expressed by P. Hence, if n denotes no object, nP must be false. Thus, if there is no unique thing having the property P (if, for example, P is 'is a winged horse'), $(\iota x\, xP)P$ is false. As is to be expected, under these conditions, the CP may fail.

Now, how does all this bear on the Ontological Argument? Recall that the instance of the CP invoked there is $\gamma P_1 \& \ldots \& \gamma P_n$, where γ is the description $\iota x(xP_1 \& \ldots \& xP_n)$. Either there is something satisfying $xP_1 \& \ldots \& xP_n$, or there is not. If there is, it must be unique. (There cannot be two omnipotent objects: if I am omnipotent, I can stop you doing things, so you cannot be omnipotent.) So γ refers to this thing, and $\gamma P_1 \& \ldots \& \gamma P_n$ is true. If there is not, then γ refers to nothing; so each conjunct of γP_1 &...& γP_n is false; as, therefore, is the whole conjunction. In other words, the instance of the CP used in the argument is true enough if God exists; but it is false if God does not exist. So if one is arguing for the existence of God, one cannot simply invoke this instance of the CP: that would just be *assuming* what one is supposed to be proving. Philosophers say that such an argument *begs the question*; that is, begs to be granted exactly what is in question. And an argument that begs the question clearly does not work.

So much for the Ontological Argument. Let us finish this chapter by seeing that the account of descriptions that I have explained is itself problematic in certain ways. According to this account, if δP is a sentence where δ is a description that does not refer to anything, it is false. But this does not always seem to be right. For example, it would seem to be true that the most powerful of the ancient Greek gods was called 'Zeus', lived on Mount Olympus, was worshipped by the Greeks, and so on. Yet there were, in reality, no ancient Greek gods. They did not, in fact, exist. If this is right, then the description 'the most powerful of the ancient Greek gods' does not refer to anything. But in that case, there are true subject/predicate sentences in which the subject term fails to refer to anything, such as 'The most powerful of the ancient Greek gods was worshipped by the Greeks'. To put it tendentiously, there are truths about non-existent objects, after all.

Main idea of the chapter

- $(\iota x c_x)P$ is true in a situation just if, in that situation, there is a unique object, *a*, satisfying c_x, and *aP*.

Chapter 5
Self-reference: what is this chapter about?

Often, things seem simple when one thinks about normal cases; but this can be deceptive. When one considers more unusual cases, the simplicity may well disappear. So it is with reference. We saw in Chapter 4 that things are not as straightforward as one might have supposed, once one takes into account the fact that some names may not refer to anything. Further complexities arise when we consider another kind of unusual case: self-reference.

It is quite possible for a name to refer to something of which it, itself, is part. For example, consider the sentence 'This sentence contains five words'. The name which is the subject of this sentence, 'this sentence', refers to the whole sentence, of which that name is a part. Similar things happen in a set of regulations which contain the clause 'These regulations may be revised by a majority decision of the Department of Philosophy', or by a person who thinks 'If I am thinking this thought, then I must be conscious'.

These are all relatively unproblematic cases of self-reference. There are other cases which are quite different. For example, suppose someone says:

> This very sentence that I am now uttering is false.

Call this sentence λ. Is λ true or false? Well, if it is true, then what it says is the case, so λ is false. But if it is false, then, since this is exactly what it claims, it is true. In either case, λ would seem to be both true and false. The sentence is like a Möbius strip (Figure 5), a topological configuration where, because of a twist, the inside is the outside, and the outside is the inside: truth is falsity, and falsity is truth.

Or suppose someone says:

> This very sentence that I am now uttering is true.

Is that true or false? Well, if it is true, it is true, since that is what it says. And if it is false, then it is false, since it says that it is true.

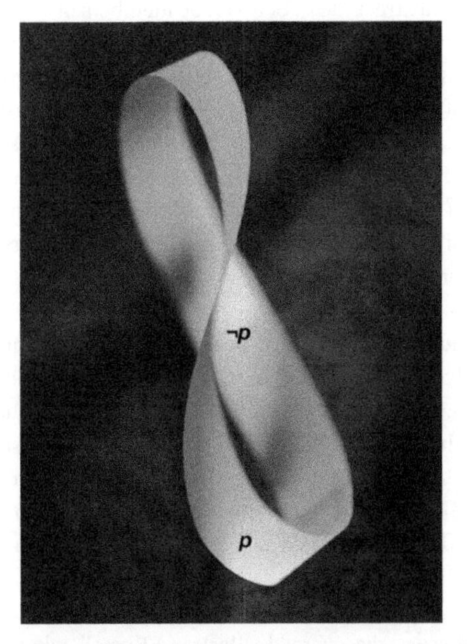

5. A Möbius strip. Inside is outside, and outside in. Truth is falsity, and falsity truth.

Hence, both the assumption that it is true and the assumption that it is false appear to be consistent. Moreover, there would seem to be no other fact that settles the matter of what truth value it has. It's not just that it has some value which we don't, or even can't, know. Rather, there would seem to be nothing that determines it as either true or false at all. It would seem to be neither true nor false.

These paradoxes are very ancient. The first of them appears to have been discovered by the ancient Greek philosopher Eubulides, and is often called the *liar paradox*. There are many more, and more recent, paradoxes of the same kind, some of which play a crucial role in central parts of mathematical reasoning. Here is another example. A set is a collection of objects. Thus, for example, one may have the set of all people, the set of all numbers, the set of all abstract ideas. Sets can be members of other sets. Thus, for example, the set of all the people in a room is a set, and hence is a member of the set of all sets. Some sets can even be members of themselves: the set of all the objects mentioned on this page is an object mentioned on this page (I have just mentioned it), and so a member of itself; the set of all sets is a set, and so a member of itself. And some sets are certainly not members of themselves: the set of all people is not a person, and so not a member of the set of all people.

Now, consider the set of all those sets that are not members of themselves. Call this R. Is R a member of itself, or is it not? If it is a member of itself, then it is one of the things that is not a member of itself, and so it is not a member of itself. If, on the other hand, it is not a member of itself, it is one of those sets that are not members of themselves, and so it *is* a member of itself. It would seem that R both is and is not a member of itself.

This paradox was discovered by Bertrand Russell, whom we met in Chapter 4, and so is called *Russell's paradox*. Like the liar paradox, it has a cousin. What about the set of all sets that *are* members of themselves. Is this a member of itself, or is it not? Well, if it is, it is; and if it is not, it is not. Again, there would seem to be nothing to determine the matter either way.

What examples of this kind do is challenge the assumption we made in Chapter 2—that every sentence is either true or false, but not both. 'This sentence is false', and 'R is not a member of itself' seem to be both true and false; and their cousins seem to be neither true nor false.

How can this idea be accommodated? Simply by taking these other possibilities into account. Assume that in any situation, every sentence is true but not false, false but not true, both true and false, or neither true nor false. Recall from Chapter 2 that the truth conditions for negation, conjunction, and disjunction are the following. In any situation:

$\neg a$ has the value T just if a has the value F.
$\neg a$ has the value F just if a has the value T.

$a \,\&\, b$ has the value T just if both a and b have the value T.
$a \,\&\, b$ has the value F just if at least one of a and b has the value F.

$a \vee b$ has the value T just if at least one of a and b has the value T.
$a \vee b$ has the value F just if both a and b have the value F.

Using this information, it is easy to work out the truth values of complex sentences under the new regime. Some examples:

- Suppose we have the following situation:

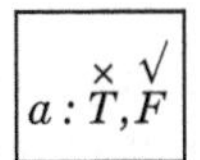

(The tick means that a has the value; a cross means that it doesn't.) Then $\neg a$ has the value T (by the first clause for negation), and $\neg a$ does not have the value F (by the second).

- Or in the following:

$$\begin{array}{|c|}\hline a : \overset{\surd}{T}, \overset{\surd}{F} \\ b : \overset{\surd}{T}, \overset{\times}{F} \\ \hline\end{array}$$

a & b has the value T (by the first clause for conjunction), and a & b has the value F (by the second).

- And in the following:

$$\begin{array}{|c|}\hline a : \overset{\surd}{T}, \overset{\times}{F} \\ b : \overset{\times}{T}, \overset{\times}{F} \\ \hline\end{array}$$

$a \vee b$ has the value T (by the first clause for disjunction), and $a \vee b$ does not have the value F (by the second).

What does this tell us about validity? A valid argument is still one where there is no situation where the premisses are true, and the conclusion is not true. And a situation is still something that gives a truth value to each relevant sentence. Only now, the situation may give a sentence one truth value, two, or none. So consider the inference $q / q \vee p$. In any situation where q has the value T, the conditions for $\vee$ assure us that $q \vee p$ also has the value T. (It may have the value F also, but no matter.) Thus, if the premiss has the value T, so does the conclusion. The inference is valid.

At this point, it is worth returning to the inference with which we started in Chapter 2: $q, \neg q / p$. As we saw in that chapter, given the assumptions made there, this inference is valid. But given the new assumptions, things are different. To see why, just take a situation where q has the values T and F, but p has just the value F. Since q is both T and F, $\neg q$ is also both T and F. Hence, both premisses are T (and F as well, but that is not relevant), and the conclusion, p, is *not* T. This gives us another diagnosis of why we find the inference intuitively invalid. It *is* invalid.

That's not the end of the matter, though. As we saw in Chapter 2, this inference follows from two other inferences. The first of these ($q / q \vee p$) we have just seen to be valid on the present account. The other must therefore be invalid; and so it is. The other inference is:

$$\frac{q \vee p \quad \neg q}{p}$$

Now consider a situation where q gets the values T and F, and p gets just the value F. It is easy enough to check that both premisses get the value T (as well as F). But the conclusion does not get the value T. Hence, the inference is invalid.

In Chapter 2, I said that this inference does seem intuitively valid. So, given the new account, our intuitions about this must be wrong. One can offer an explanation of this fact, however. The inference appears to be valid because, if $\neg q$ is true, this seems to rule out the truth of q, leaving us with p. But on the present account, the truth of $\neg q$ does not rule out that of q. It would do so only if something could not be both true and false. When we think the inference to be valid, we are perhaps forgetting such possibilities, which can arise in unusual cases, like those which are provided by self-reference.

Which explanation of the situation is better, the one that we ended up with in Chapter 2, or the one we now have? That is a question which I will leave you to think about. Let us end, instead, by noting that, as always, one may challenge some of the ideas on which the new account rests. Consider the liar paradox and its cousin. Take the latter first. The sentence 'This sentence is true' was supposed to be an example of something that is neither true nor false. Let us suppose that this is so. Then, in particular, it is not true. But it, itself, says that it *is* true. So it must be false, contrary to our supposition that it is neither true nor false. We seem to have ended up in a contradiction. Or take the liar sentence, 'This sentence is false'. This was supposed to be an example of a sentence that is both true and false. Let's tweak it a bit. Consider, instead, the sentence 'This sentence is not true'. What is the truth value of this? If it is true, then what it says is the case; so it is not true. But if it's not true, then, since that is what it says, it is true. Either way, it would seem to be both true and not true. Again, we have a contradiction on our hands. It's not just that a sentence may take the values T and F; rather, a sentence can both be T and not be T.

It is situations of this kind that have made the subject of self-reference a contentious one, ever since Eubulides. It is, indeed, a very tangled issue.

Main idea of the chapter

- Sentences may be true, false, both, or neither.

Final comment: to avoid possible misunderstandings, I note that in the chapters to follow I am going to resume the assumption that statements are true or false, but not both or neither—except in Chapter 10 (on fuzzy logic).

Chapter 6
Necessity and possibility: what will be must be?

We often claim not just that something *is* so, but that it *must be* so. We say: 'It must be going to rain', 'It can't fail to rain', 'Necessarily, it's going to rain'. We also have many ways of saying that, though something may, in fact, not be the case, it *could* be. We say: 'It could rain tomorrow', 'It is possible that it will rain tomorrow', 'It's not impossible that it will rain tomorrow'. If a is any sentence, logicians usually write the claim that a must be true as $\Box a$, and the claim that a could be true as $\Diamond a$.

$\Box$ and $\Diamond$ are called *modal operators*, since they express the modes with which things are true or false (necessarily, possibly). The two operators are, in fact, connected. To say that something must be the case is to say that it is not possible for it not to be the case. That is, $\Box a$ means the same as $\neg\Diamond\neg a$. Similarly, to say that it is possible for something to be the case is to say that it is not necessarily the case that it is false. That is, $\Diamond a$ means the same as $\neg\Box\neg a$. For good measure, we can express the fact that it is impossible for a to be true, indifferently, as $\neg\Diamond a$ (it is not possible that a), or as $\Box\neg a$ (a is necessarily false).

Unlike the operators we have met so far, $\Box$ and $\Diamond$ are not truth functions. As we saw in Chapter 2, when you know the truth value of a, you can work out the truth value of $\neg a$. Similarly, when you know the truth values of a and b, you can work out the truth values of $a \vee b$ and a & b. But you cannot infer the truth value of $\Diamond a$ simply from knowledge of the truth value of a. For example, let r be the sentence 'I will rise before 7 a.m. tomorrow': r is, as a matter of fact, false. But it certainly could be true: I could set my alarm clock and rise early. Hence, $\Diamond r$ is true. By contrast, let j be the sentence 'I will jump out of bed and hover 2m above the ground'. Like r, this is false too. But unlike r, it is not even possible that it is true. That would violate the laws of gravity. Hence, $\Diamond j$ is false. So the truth value of a sentence, a, does not determine that of $\Diamond a$: r and j are both false, but $\Diamond r$ is true and $\Diamond j$ is false. Similarly, the truth value of a does not determine the truth value of $\Box a$. Let r now be the sentence 'I will rise before 8 a.m. tomorrow'. This is, in fact, true; but it is not necessarily true. I could stay in bed. Let j now be the sentence 'If I jump out of bed tomorrow morning, I will have moved'. That is also true, but there is no way that *that* could be false. It's necessarily true. Hence, r and j are both true, but one is necessarily true, and the other is not.

Modal operators are therefore operators of a kind quite different from anything that we have met so far. They are also important and often puzzling operators. To illustrate this, here is an argument for fatalism, given by the other of the two most influential ancient Greek philosophers, Aristotle (Figure 6).

Fatalism is the view that whatever happens *must* happen: it could not have been avoided. When an accident occurs, or a person dies, there is nothing that could have been done to prevent it. Fatalism is a view that has appealed to some. When something goes wrong, there is a certain amount of comfort to be derived from the thought that it could not have been otherwise. None the less, fatalism entails that I am powerless to alter what happens, and this seems plainly false. If I am involved in a traffic accident today, I could have avoided it simply by taking a different route. So what is Aristotle's argument? It goes like this. (Ignore the boldface for the present; we will come back to this.)

6. Aristotle (384–322 BCE), the founder of formal logic.

Take any claim you like—say, for the sake of illustration, that I will be involved in a traffic accident tomorrow. Now, we may not know yet whether or not this is true, but we know that either I will be involved in an accident or I won't. Suppose the first of these. Then, as a matter of fact, I will be involved in a traffic accident. And **if it is true to say that I will be involved in an accident then it cannot fail to be the case that I will be involved**. That is, it must be the case that I will be involved. Suppose, on the other hand, that I will not, as a matter of fact, be involved in a traffic accident tomorrow. Then it is true to say that I will not be involved in an accident; and if this is so, it cannot fail to be the case that I won't be in an accident. That is, it must be the case that I am not involved in an accident. Whichever of these two *does* happen, then, it *must* happen. This is fatalism.

What is one to say about this? To answer this, let us look at a standard modern understanding of modal operators. We suppose that every situation, s, comes furnished with a bunch of possibilities, that is, situations that are possible as far as s goes—to be definite, let us say situations that could arise without violating the laws of physics. Thus, if s is the situation that I am presently in (being in Australia), my being in London in a week's time is a possible situation; whilst my being on Alpha Centauri (over four light-years away) is not. Following the 17th-century philosopher and logician Leibniz, logicians often call these possible situations, colourfully, *possible worlds*. Now, to say that $\Diamond a$ (it is possibly the case that a) is true in s, is just to say that a is in fact true in *at least one* of the possible worlds associated with s. And to say that $\Box a$ (it is necessarily the case that a) is true in s, is just to say that a is true in *all* the possible worlds associated with s. This is why $\Box$ and $\Diamond$ are not truth functions. For a and b may have the same truth value in s, say F, but may have different truth values in the worlds associated with s. For example, a may be true in one of them (say, s'), but b may be true in none, like this:

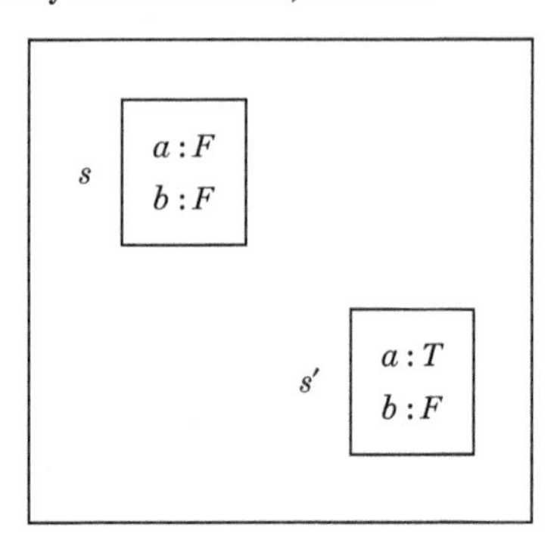

This account gives us a way of analysing inferences employing modal operators. For example, consider the inference:

$$\frac{\Diamond a \quad \Diamond b}{\Diamond(a \,\&\, b)}$$

This is invalid. To see why, suppose that the situations associated with s are s_1 and s_2, and that truth values are as follows:

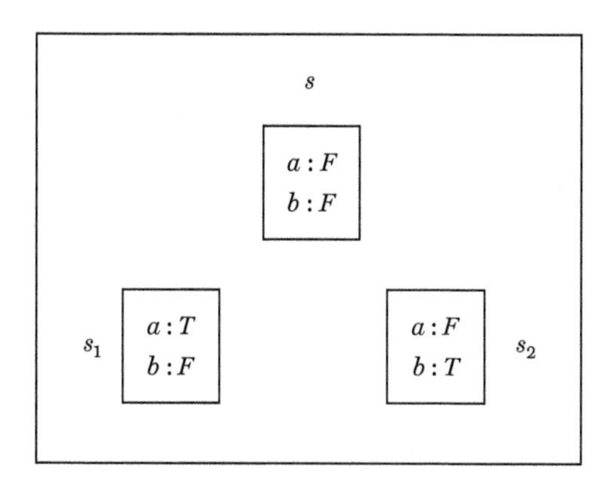

a is T at s_1; hence, $\Diamond a$ is true in s. Similarly, b is true in s_2; hence, $\Diamond b$ is true in s. But a & b is true in no associated world; hence, $\Diamond(a \,\&\, b)$ is not true in s.

By contrast, the following inference is valid:

$$\frac{\Box a \quad \Box b}{\Box(a \,\&\, b)}$$

For if the premisses are true in a situation s, then a and b are true in all the worlds associated with s. But then a & b is true in all those worlds. That is, $\Box(a \,\&\, b)$ is true in s.

Before we can get back to the question as to how this bears on Aristotle's argument, we need to talk briefly about another logical operator that we have not yet met. Let us write 'if a then b' as $a \rightarrow b$. Sentences of this form are called *conditionals*, and will concern us a good deal in Chapter 7. All we need to note for the present is that the major inference that conditionals seem to be involved in is this:

$$\frac{a \quad a \to b}{b}$$

(For example: 'If she works out regularly then she is fit. She does work out regularly; so she is fit'.) Modern logicians usually call this inference by the name with which it was tagged by medieval logicians: *modus ponens*. Literally, this means 'the method of positing'. (Don't ask.)

Now, for Aristotle's argument, we need to think a little about conditionals of the form:

> if a then it cannot fail to be the case that b.

Such sentences are, in fact, ambiguous. One thing they can mean is that if a is, as a matter of fact, true, then b is necessarily true. That is, if a is true in the situation we are talking about, s, then b is true in all the possible situations associated with s. We can write this as $a \to \Box b$. The sentence is being used like this when we say things like: 'You can't change the past. If something is true of the past, it cannot now fail to be true. There is nothing you can do to make it otherwise: it's irrevocable.'

The second meaning of a conditional of the form 'If a then it cannot fail to be the case that b', is quite different. We often use this form of words to express the fact that b follows from a. We would be using the sentence like this if we said something like 'If Fred is going to be divorced then he cannot fail to be married'. We are not saying that if Fred is going to be divorced, his marriage is irrevocable. We are saying that you can't get a divorce unless you are married. There is no possible situation in which you have the one, but not the other. That is, in any possible situation, if one is true, so is the other. That is, $\Box(a \to b)$ is true.

Now, $a \to \Box b$ and $\Box(a \to b)$ mean quite different things. And certainly, the first does not follow from the second. The mere fact that $a \to b$ is true in every situation associated with s, does not mean that $a \to \Box b$ is true in s; a could be true in s, whilst $\Box b$ is not—both b *and* a may fail to be true in some associated world. Or, to give a concrete counter-example: it is necessarily true that if John is getting a divorce, he is married; but it is certainly not true that if John is getting a divorce he is necessarily (irrevocably) married.

To come back to Aristotle's argument at last, consider the sentence that I put in boldface: 'if it is true to say that I will be involved in an accident then it cannot fail to be the case that I will be involved'. This is exactly of the form we have just been talking about. It is therefore ambiguous. Moreover, the argument trades on this ambiguity. If a is the sentence 'It is true to say that I will be involved in a traffic accident', and b is the sentence 'I will be involved (in a traffic accident)', then the boldface conditional is true in the sense:

1 $\Box(a \to b)$

Necessarily, if it is true to say something, then that something is indeed the case. But what needs to be established is:

2 $a \to \Box b$

After all, the next step of the argument is precisely to infer $\Box b$ from a by *modus ponens*. But as we have seen, 2 does not follow from 1 at all. Hence, Aristotle's argument is invalid. For good measure, exactly the same problem arises in the second part of the argument, with the conditional 'if it is true to say that I will not be involved in an accident then it cannot fail to be the case that I won't be involved in an accident'.

This seems a satisfactory reply to Aristotle's argument. But there is a closely related argument that cannot be answered so easily. Come back to the example we had about changing the past. It does seem to be true that if some statement about the past is true, it is now necessarily true. It is impossible, now, to render it false. The Battle of Hastings was fought in 1066, and there is now nothing that one can do to make it have been fought in 1067. Thus, if p is some statement about the past, $p \rightarrow \Box p$.

Now consider some statement about the future. Again, for example, let it be the claim that I will be involved in a traffic accident tomorrow. Suppose this is true. Then if someone uttered this sentence 100 years ago, they spoke truly. And even if no one actually uttered it, if anyone had uttered it, they would have spoken truly. Thus, that I will be involved in a traffic accident tomorrow was true 100 years ago. This statement (p) is certainly a statement about the past, and so, since true, necessarily true ($\Box p$). So it must necessarily be true that I will be involved in a traffic accident tomorrow. But that was just an example; the same reasoning could be applied to anything. Thus, anything that happens, must happen. This argument for fatalism does not commit the same fallacy (that is, use the same invalid argument) as the first one that I gave. So is fatalism true after all?

Main ideas of the chapter

- Each situation comes with a collection of associated possible situations.
- $\Box a$ is true in a situation, s, if a is true in every situation associated with s.
- $\Diamond a$ is true in a situation, s, if a is true in some situation associated with s.

Chapter 7
Conditionals: what's in an *if*?

In this chapter we'll turn to the logical operator that I introduced in passing in Chapter 6, the conditional. Recall that a conditional is a sentence of the form 'if a then c', which we are writing as $a \to c$. Logicians call a the *antecedent* of the conditional, and c the *consequent*. We also noted that one of the most fundamental inferences concerning the conditional is *modus ponens*: $a, a \to c / c$. Conditionals are fundamental to much of our reasoning. Chapter 6 showed just one example of this. Yet they are deeply puzzling. They have been studied in logic ever since its earliest times. In fact, it was reported by one ancient commentator (Callimachus) that at one time even the crows on the rooftops were cawing about conditionals.

Let us see why—or, at least, one reason why—conditionals are puzzling. If you know that $a \to c$, it would seem that you can infer that $\neg(a \,\&\, \neg c)$ (it is not the case that a and not c). Suppose, for example, that someone informs you that if you miss the bus, you will be late. You can infer from this that it is false that you will miss the bus and not be late. Conversely, if you know that $\neg(a \,\&\, \neg c)$, it would seem that you can infer $a \to c$ from this. Suppose, for example, that someone tells you that you won't go to the movies without spending money (it's not the case that you go to the movies and do not spend money). You can infer that if you go to the movies, you will spend money.

$\neg(a \& \neg c)$ is often written as $a \supset c$, and called the *material conditional*. Thus, it would appear that $a \rightarrow c$ and $a \supset c$ mean much the same thing. In particular, assuming the machinery of Chapter 2, they must have the same truth table. It is a simple exercise, which I leave to you, to show that this is as follows:

a	c	$a \supset c$
T	T	T
T	F	F
F	T	T
F	F	T

But this is odd. It means that if c is true in a situation (first and third rows), so is $a \rightarrow c$. This hardly seems right. It is true, for example, that Canberra is the federal capital of Australia, but the conditional 'If Canberra is not the federal capital of Australia, Canberra is the federal capital of Australia' seems plainly false. Similarly, the truth table shows us that if a is false (third and fourth rows), $a \rightarrow c$ is true. But this hardly seems right either. The conditional 'If Sydney is the federal capital of Australia, then Brisbane is the federal capital' also appears patently false. What has gone wrong?

What these examples seem to show is that $\rightarrow$ is not a truth function: the truth value of $a \rightarrow c$ is not determined by the truth values of a and c. Both of 'Rome is in France' and 'Beijing is in France' are false; but it's true that:

If Italy is part of France, Rome is in France.

While it's false that:

If Italy is part of France, Beijing is in France.

So how do conditionals work?

One answer can be given using the machinery of possible worlds of Chapter 6. Consider the last two conditionals. In any possible situation in which Italy had become incorporated into France, Rome would indeed have been in France. But there are possible situations in which Italy was incorporated in France, but this had no effect on China at all. So Beijing was still not in France. This suggests that the conditional $a \rightarrow c$ is true in some situation, s, just if c is true in every one of the possible situations associated with s in which a is true; and it is false in s if c is false in some possible situation associated with s in which a is true.

This gives a plausible account of $\rightarrow$. For example, it shows why *modus ponens* is valid—at least on one assumption. The assumption is that we count s itself as one of the possible situations associated with s. This seems reasonable: anything that is *actually* the case in s is surely *possible*. Now, suppose that a and $a \rightarrow c$ are true in some situation, s. Then c is true in all situations associated with s in which a is true. But s is one of those situations, and a is true in it. Hence, so is c, as required.

Going back to the argument with which we started, we can now see why it fails. The inference on which the argument depends is:

$$\frac{\neg(a \,\&\, \neg c)}{a \rightarrow c}$$

And this is not valid. For example, if a is F in some situation, s, this suffices to make the premiss true in s. But this tells us nothing about how a and c behave in the possible situations associated with s. It could well be that in one of these, say s', a is true and c is not, like this:

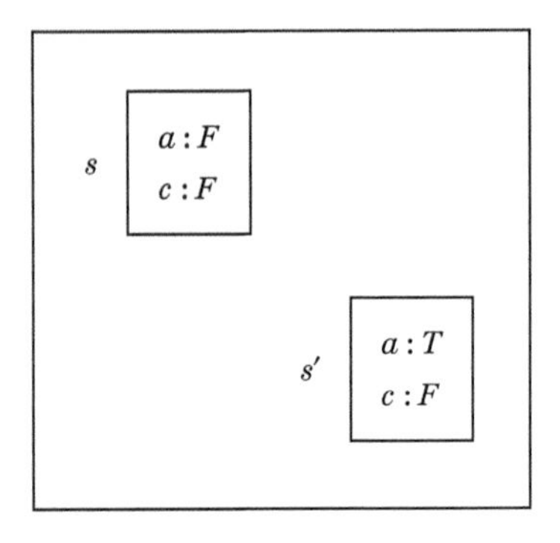

So $a \rightarrow c$ is not true at s.

What about the example we had earlier, where you are informed that you won't go to the movies without spending money? Didn't the inference seem valid in this case? Suppose you know that you won't go to the movies without spending money: $\neg(g \,\&\, \neg m)$. Are you really entitled to conclude that if you go to the movies you will spend money: $g \rightarrow m$? Not necessarily. Suppose you are not going to go to the movies, come what may, even if admission is free that night. (There is a programme on the television that is much more interesting.) Then you know that it is not true that you will go ($\neg g$), and so that it is not true that you will go *and* not spend money: $\neg(g \,\&\, \neg m)$. Are you then entitled to infer that if you go you will spend money? Certainly not: it may be a free night.

It is important to note that in the kind of situation where you learn that the premiss is true by being informed of it, other factors are usually operating. When someone tells you something like $\neg(g \,\&\, \neg m)$, they do not normally do this on the basis that they know that $\neg g$ is true. (If they knew this, there wouldn't normally be a point in telling you anything much about the situation.) If they tell you this, it is on the basis that there is some connection between g and m: that you *can't* have g true without m being true—and that is exactly what it takes for the conditional to be true. So in the case where you are informed of the premiss, it would normally be reasonable to infer $g \rightarrow m$; but not from the content of what was said—rather, from the fact *that it was said.*

In fact, we often correctly make inferences of this kind without thinking. Suppose, for example that I ask someone how to get my computer to do something or other, and they reply 'There is a manual on the shelf'. I infer that it is a computer manual. This does not follow from what was actually said, but the remark would not have been relevant unless the manual was a computer manual, and people are normally relevant in what they say. Hence, I can conclude that it is a computer manual from the fact that they said what they did. The inference is not deductively valid. After all, the person could have said this, and it not be a computer manual (Figure 7). But the inference is still an excellent inductive inference. It is of a kind usually called *conversational implicature.*

The account of the conditional that we have just been looking at seems to fare well—at least as far as we have looked. It faces a number of problems, though. Here is one. Consider the following inferences:

> If you go to Rome you will be in Italy.
> If you are in Italy, you are in Europe.
> Hence, if you go to Rome, you will be in Europe.

> If x is greater than 10 then x is greater than 5.
> Hence, if x is greater than 10 and less than 100, then x is greater than 5.

7. **Jumping to conclusions.**

These inferences seem perfectly valid, and so they are on the present account. We can write the first inference as:

1 $$\frac{a \rightarrow b \quad b \rightarrow c}{a \rightarrow c}$$

To see that this comes out valid, suppose the premisses are true in some situation, *s*. Then *b* is true in every possible situation associated with *s* where *a* is true; and likewise, *c* is true in every associated situation where *b* is. So *c* is true in every such situation where *a* is true. That is, $a \rightarrow c$ is true in *s*.

We can write the second inference as:

$$\mathbf{2} \quad \frac{a \rightarrow c}{(a \,\&\, b) \rightarrow c}$$

To see that this comes out valid, suppose the premiss is true in some situation, s. Then c is true in every possible situation associated with s where a is true. Now, suppose a & b is true in an associated situation; then a is certainly true in that situation, and hence c is. Hence, $(a \,\&\, b) \rightarrow c$ is true in s.

So far so good. The problem is that there are inferences that are exactly of these forms, but which appear to be *invalid*. For example, suppose that there is an election for Prime Minister with only two candidates, Smith, the present Prime Minister, and Jones. Now consider the following inference:

> If Smith dies before the election, Jones will win. If Jones wins the election, Smith will retire and take her pension. Hence, if Smith dies before the election, she will retire and take her pension.

This is exactly an inference of the form 1. But it seems clear that there could be a situation in which both premisses are true. But not the conclusion—unless we are considering a bizarre situation in which the government can effect pension payments in the after-life!

Or consider the following inference concerning said Smith:

> If Smith jumps from the top of a tall precipice, she will die from the fall. Hence, if Smith jumps from the top of a tall precipice and wears a parachute, she will die from the fall.

This is an inference of the form 2. Yet, again, it would seem clear that there could be situations where the premiss is true and the conclusion is not.

What is one to say about this state of affairs? I'll leave you to think about that. Despite the fact that conditionals are central to how we reason about most things, they are still one of the most contentious areas of logic. If birds are no longer crowing about conditionals, logicians certainly are.

Main idea of the chapter

- $a \to b$ is true in a situation, s, just if b is true in every situation associated with s where a is true.

Chapter 8
The future and the past: is time real?

Time is something that we are all very familiar with. We plan to do things in the future; we remember things in the past; and sometimes we enjoy just being in the present. And part of our finding our way around in time is making inferences that concern time. For example, the two following inferences are intuitively valid:

$$\frac{\text{It is raining.}}{\text{It will have been raining.}} \qquad \frac{\text{It will be true that it has always been raining.}}{\text{It is raining.}}$$

All this seems elementary.

But as soon as one starts to think about time, one seems to get tangled in knots. As Augustine said, if no one asks me what time is, then I know very well; but when someone asks me, I cease to know. One of the most puzzling things about time is that it seems to flow. The present seems to move: first it is today; then it is tomorrow; and so on. But how can time change? Time is what measures the rate at which *everything else* changes. This problem is at the heart of several conundrums concerning time. One such was put forward, early in the 20th century, by the British philosopher John McTaggart Ellis McTaggart. (That's right.) Like many philosophers, McTaggart was tempted by the view that time is unreal—that, in the ultimate order of things, time is an illusion.

To explain McTaggart's argument for this, it will help to have a little symbolism. Take a past-tense sentence, such as 'The sun was shining'. We can express this equivalently, if a little awkwardly, as 'It was the case that the sun is shining'. Let us write 'It was the case that' as **P** (for 'past'). Then we can write this sentence as '**P** the sun is shining', or, writing *s* for 'The sun is shining', simply **P***s*. Similarly, take any future tense sentence, say, 'The sun will be shining'. (Strictly speaking, grammarians will tell you, English has no proper future tense, unlike French or Latin. But you know what I mean.) We can write this as 'It will be the case that the sun is shining'. If we write 'It will be the case that' as **F** (for 'future'), then we can write this as **F***s*. (Don't confuse this **F** with the truth value *F*.)

P and **F** are operators, like □ and ◊, that affix to whole sentences to make whole sentences. Moreover, like □ and ◊, they are not truth functions. 'It is 4 p.m.' and 'It is 4 p.m. on 2nd August 1999' are both true (at the instant I write); 'It *will be* 4 p.m.' is also true (at the present instant)—it is 4 p.m. once every day—though 'It *will be* 4 p.m., 2nd August 1999' is not. Logicians call **P** and **F** *tense operators*. Tense operators can be iterated, or compounded. For example, we can say 'The sun *will have been* shining', that is, 'It will be the case that it was the case that the sun is shining': **FP***s*. Or we can say 'The sun *had been* shining', that is, 'It was the case that it was the case that the sun is shining': **PP***s*. (The modal operators that we met in Chapter 7 can also be iterated in this way, though we did not consider this there.) Not all iterations of tense operators have snappy English expressions. For example, there is not a much better way to express **FPF***s* than as the rather lame 'It will be the case that it was the case that the sun will be shining'. The iterations, though, make perfectly good grammatical sense. We can call iterations of **P** and **F**, like **FP**, **PP**, **FFP**, *compound tenses*.

Now, back to McTaggart. McTaggart reasoned that there would be no time if there were no past and future: these are of its essence. Yet pastness and futurity, he argued, are inherently contradictory; so nothing in reality can correspond to them. Well, maybe. But why are past and future contradictory? For a start, past and future are incompatible. If some instantaneous event is past, it is not future, and vice versa. Let *e* be some instantaneous event. It can be anything you like, but let us suppose that it is the passing of the first bullet through the heart of Czar Nicholas in the Russian Revolution. Let *h* be the sentence '*e* is occurring'. Then we have:

$$\neg(\mathbf{P}h \,\&\, \mathbf{F}h)$$

But *e*, like all events, is past and future. Because time flows, all events have the property of being future (before they happen) *and* the property of being past (after they happen):

$$\mathbf{P}h \,\&\, \mathbf{F}h$$

So we have a contradiction.

This argument isn't likely to persuade anyone for very long. An event can't be past and future *at the same time*. The instant the bullet passed through the Czar's heart was past and future *at different times*. It started off as future; became present for a painful instant; and then was past. But now—and this is the cunning part of McTaggart's argument—what are we saying here? We are applying compound tenses to *h*. We are saying that it was the case that the event was future, **PF***h*; then it was the case that it was past, **PP***h*. Now, many compound tenses, like simple tenses, are incompatible. For example, if any event *will be* future, it is not the case that it *was* past:

$$\neg(\mathbf{PP}h \,\&\, \mathbf{FF}h)$$

But, just as with the simple tenses, the flow of time suffices to ensure that all events have all compound tenses too. In the past, **F***h*; so in the distant past **FF***h*. In the future, **P***h*; so in the distant future, **PP***h*:

PP*h* & **FF***h*

And we are back with a contradiction.

Those who have kept their wits about them will reply, just as before, that *h* has its compound tenses at different times. It was the case that **FF***h*; then, *later on*, it was the case that **PP***h*. But what are we saying here? We are applying more complex compound tenses to *h*: **PFF***h* and **PPP***h*; and we can run exactly the same argument again with these. These compound tenses are not all consistent with each other, but the flow of time ensures that *h* possesses all of them. We may make the same reply *again*, but it, too, is open to the same counter-reply. Whenever we try to get out of the contradiction with one set of tenses, we do so only by describing things in terms of other tenses that are equally contradictory; so we never escape contradiction. That is McTaggart's argument.

What is one to say about this? To answer this, let us look at the validity of inferences concerning tenses. To account for this, we suppose that every situation, s_0, comes together with a bunch of other situations—not, this time, situations that represent possibilities associated with s_0 (as with modal operators), but situations that are either before s_0 or after s_0. Assuming, as we normally do, that time is one-dimensional and infinite in both directions, past and future, we can represent the situations in a familiar way:

$$\ldots\ s_{-3}\quad s_{-2}\quad s_{-1}\quad s_0\quad s_1\quad s_2\quad s_3\ \ldots$$

Left is earlier; right is later. As usual, each s provides a truth value, T or F, for every sentence without tense operators. What about sentences with tense operators? Well, **P**a is T in any situation, s, just if a is true in some situation to the left of s; and **F**a is true in s just if a is true in some situation to the right of s.

While we are doing all this, we can add two new tense operators, **G** and **H**. **G** can be read 'It is always **G**oing to be the case that', and **G**a is true in any situation, s, just if a is true in *all* situations to the right of s. **H** can be read as 'It **H**as always been the case that', and **H**a is true in any situation, s, just if a is true in all situations to the left of s. (**G** and **H** correspond to **F** and **P**, respectively, in just the way □ that corresponds to ◊.)

This machinery shows us why the two inferences with which we started the chapter are valid. Employing tense operators, these inferences can be written, respectively, as:

$$\frac{r}{\mathbf{FP}r} \qquad \frac{\mathbf{FH}r}{r}$$

The first inference is valid, since if r is true in some situation, s_0, then in any situation to the right of s_0, say s_1, **P**r is true (since s_0 is to its left). But then, **FP**r is true in s_0, since s_1 is to its right. We can depict things like this:

$$\ldots s_{-3} \quad s_{-2} \quad s_{-1} \quad s_0 \quad s_1 \quad s_2 \quad s_3 \ldots$$

$$r$$

$$\mathbf{P}r$$

$$\mathbf{FP}r$$

The second inference is valid, since if **FH**r is true in s_0, then in some situation to the right of s_0, say s_2, **H**r is true. But then in all situations to the left of s_2, and so in particular s_0, r is true:

…	s_{-3}	s_{-2}	s_{-1}	s_0	s_1	s_2	s_3	…
				$\mathbf{FH}r$				
						$\mathbf{H}r$		
	r	r	r	r	r			

Moreover, certain combinations of tenses are impossible, as one would expect. Thus, if h is a sentence that is true in just one situation, say s_0, then $\mathbf{P}h$ & $\mathbf{F}h$ is false in every s. Both conjuncts are false in s_0; the first conjunct is false to the left of s_0; the second conjunct is false to the right. Similarly, e.g., $\mathbf{PP}h$ & $\mathbf{FF}h$ is false in every s. I leave you to check the details.

Now, how does all this bear on McTaggart's argument? The upshot of McTaggart's argument, recall, was that, given that h has every possible tense, it is never possible to avoid contradiction. Resolving contradictions in one level of complexity for compound tenses only creates them in another. The account of the tense operators that I have just given, shows this to be false. Suppose that h is true in just s_0. Then any statement with a compound tense concerning h is true *somewhere*. For example, consider $\mathbf{FPPF}h$. This is true in s_{-2}, as the following diagram shows:

…	s_{-3}	s_{-2}	s_{-1}	s_0	s_1	s_2	s_3	…
				h				
			$\mathbf{F}h$					
					$\mathbf{PF}h$			
						$\mathbf{PPF}h$		
		$\mathbf{FPPF}h$						

Clearly, we can do the same for every compound tense composed of **F** and **P**, zigzagging left or right, as required. And all this is perfectly consistent. The infinitude of different situations allows us to assign h all its compound tenses in appropriate places without violating the various incompatibilities between them, e.g., by having **F***h* and **P***h* true in the same situation. McTaggart's argument, therefore, fails.

This is a happy outcome for those who wish to believe in the reality of time. But those who agree with McTaggart might yet not be persuaded by our considerations. Suppose I give you a set of specifications for constructing a house: the front door goes here; a window here.... How do you know that all the specifications are consistent? How do you know that, when you perform the construction, everything will work out, and that you will not be required, for example, to put the door in incompatible positions? One way to determine this is to build a scale model in accordance with all the specifications. If such a model can be built, the specifications are consistent. That is exactly what we have done with our tensed talk. The model is the sequence of situations, together with the way of assigning *T* and *F* to tensed sentences. It is a little more abstract than a model of a house, but the principle is essentially the same.

It may be possible to object to a model, though. Sometimes a model will ignore important things. For example in a scale model of a house, a beam may not collapse, because it bears a lot less stress than the corresponding beam would in a full-scale construction. The full-scale beam may be required to take an impossible load, making the full-scale building impossible—the model notwithstanding. Similarly, it may be suggested that our model of time ignores important things. After all, what we have done is give a *spatial* model of time (left, right, etc.). But space and time are quite different things (Figure 8). Space does not flow in the way that time does (whatever, indeed, that might mean). Now, it is exactly the flow of time that produces the supposed contradiction that McTaggart was pointing to. No wonder this does not show up in the model! Exactly what, then, is missing from the model? And once that is taken into account, does the contradiction reappear?

8. For Dr Who, anyway, space and time are much the same thing.

Main ideas of the chapter

- Every situation comes with an associated collection of earlier and later situations.
- **F**a is true in a situation if a is true in some later situation.
- **P**a is true in a situation if a is true in some earlier situation.
- **G**a is true in a situation if a is true in every later situation.
- **H**a is true in a situation if a is true in every earlier situation.

Chapter 9
Identity and change: is anything ever the same?

We have not finished with time yet. Time is involved in various other conundrums, one kind of which we will look at in this chapter. This kind concerns problems that arise when things change; and specifically, the question of what is to be said about the identity of objects that change through time.

Here is an example. We all think that objects can survive through change. For example, when I paint a cupboard, although its colour may change, it is still the same cupboard. Or when you change your hairstyle, or if you are unfortunate enough to lose a limb, you are still *you*. But how can anything survive change? After all, when you change your hairstyle, the person that results is different, not the same at all. And if the person is different, it is a different person; so the old you has gone out of existence. In exactly the same way, it may be argued, no object persists through any change whatsoever. For any change means that the old object goes out of existence, and is replaced by a quite different object.

Arguments like this appear at various places in the history of philosophy, but it would be generally agreed by logicians, now, that they are mistaken, and rest on a simple ambiguity. We must distinguish between an object and its properties. When we say that you, with a different hairstyle, are different, we are saying that you have different properties. It does not follow that you are literally a different person, in the way that I am a different person from you.

One reason why one may fail to distinguish between being a certain object and having certain properties is that, in English, the verb 'to be' and its various grammatical forms—'is', 'am', and so on—can be used to express both of these things. (And the same goes for similar words in other languages.) If we say 'The table is red', 'Your hair is now short', and similar things, we are attributing a property to an object. But if someone says 'I am Graham Priest', 'The person who won the race is the same person who won it last year', and so on, then they are identifying an object in a certain way. That is, they are stating its identity.

Logicians call the first use of 'is' the *'is' of predication*; they call the second use of 'is' the *'is' of identity*. And because these have somewhat different properties, they write them in different ways. The 'is' of predication we have already met in Chapter 3. 'John is red' is typically written in the form jR, so the 'is' is incorporated in the predicate. (Actually, as I noted in Chapter 3, it is more common to write this the other way round, as Rj.) The 'is' of identity is written with =, familiar from school mathematics. Thus, 'John is the person who won the race' is written: $j = w$. (The name w is a description here; but this is of no significance in the present matter.) Sentences like this are called *identities*.

What properties does identity have? First, it is a relation. A relation is something that relates two objects. For example, *seeing* is a relation. If we say 'John sees Mary' we are stating a relation between them. The objects related by a relation do not necessarily have to be different. If we say 'John sees himself' (maybe in a mirror), we are stating a relation that John bears to John. Now, identity is a very special relation. It is a relation that every object bears to itself and to nothing else.

You might think that this would make identity a rather useless relation, but, in fact, this is not so. For example, if I say 'John is the person who won the race', I am saying that the relation of identity holds between the object referred to by 'John' and the object referred to by 'the person who won the race'—in other words, that these two names refer to one and the same person. This can be a highly significant piece of information.

The most important things about identity, though, are the inferences that it is involved in. Here is an example:

> John is the person who won the race.
> The person who won the race got a prize.
> So John got a prize.

We can write this as:

$$\frac{j = w \quad wP}{jP}$$

This inference is valid in virtue of the fact that, for any objects, x and y, if $x = y$, then x has any properties that y has, and vice versa. One and the same object either has the property in question, or it doesn't. This is usually called *Leibniz's Law*, after Leibniz (Figure 9), whom we met in Chapter 6. In an application of Leibniz's Law, one premiss is an identity statement, say $m = n$; the second premiss is a sentence containing one of the names that flanks the identity sign, say m; and the conclusion is obtained by substituting n for m in this.

Leibniz's Law is a very important one, and has many quite unproblematic applications. For example, high school algebra assures us that $(x+y)(x-y)= x^2 - y^2$. So if you are solving a problem, and establish that, say, $x^2 - y^2 = 0$, you can apply Leibniz's Law to infer that $(x+y)(x-y)=0$. Its deceptive simplicity hides a multitude of problems, though. In particular,

9. Gottfried Wilhelm Leibniz (1646–1716), the last notable logician before the modern period.

there seem to be many counter-examples to it. Consider, for example, the following inference:

> John is the person who won the race.
> Mary knows that the person who won the race got a prize.
> So Mary knows that John got a prize.

This looks like an application of Leibniz's Law since the conclusion is obtained by substituting 'John' for 'the person who won the race' in the second premiss. Yet it is clear that the premisses could well be true without the conclusion being true: Mary might not know that John is the person who won the race. Is this a violation of Leibniz's Law? Not necessarily. The Law says that if $x = y$ then any property of x is a property of y. Now, does the condition

'Mary knows that x got a prize' express a property of x? Not really: it would seem, rather, to express a property of Mary. If Mary were suddenly to go out of existence, this would not change x at all! (The logic of phrases such as 'knows that' is still very much *sub judice* in logic.)

Another sort of problem is as follows. Here is a road; it is a tarmac road; call it t. And here is a road; it is a dirt road; call it d. The two roads, though, are the same road, $t = d$. It is just that the tarmac runs out towards the end of the road. So Leibniz's Law tells us that t is a dirt road, and d is a tarmac road—which they are not. What has gone wrong here? We cannot say that being dirt or tarmac are not really properties of the road. They certainly are. What has gone wrong (arguably) is this: we are not being precise enough in our specification of properties. The relevant properties are *being tarmac at such and such a point*, and *being dirt at such and such a point*. Since t and d are the same road, they have both properties, and we do not have a violation of Leibniz's Law.

So far so good. These problems are relatively easy. Let's now have one that isn't. And here, time comes back into the issue. To explain what the problem is, it will be useful to employ the tense operators of Chapter 8, and specifically, **G** ('it is always going to be the case that'). Let x be anything you like, a tree, a person; and consider the statement $x = x$. This says that x has the property of being identical to x—which is obviously true: it's part of the very meaning of identity. And this is so, regardless of time. It is true now, true at all times future, and true at all times past. In particular, then, $\mathbf{G}\, x = x$ is true. Now, here is an instance of Leibniz's Law:

$$\frac{x = y \quad \mathbf{G}\, x = x}{\mathbf{G}\, x = y}$$

(Don't let the fact that we have substituted y for only one of the occurrences of x in the second premiss throw you. Such applications of Leibniz's Law make perfectly good sense.

Just consider: 'John is the person who won the race; John sees John; so John sees the person who won the race'.) What the inference shows is that if x is identical to y, and x has the property of being identical to x at all future times, so does y. And since the second premiss is true, as we have just noted, it follows that if two things are identical, they will always be identical.

And what of that? Simply, it doesn't always seem to be true. For example, consider an amoeba. Amoebas are single-celled water creatures that multiply by fission: an amoeba will split down the middle to become two amoebas. Now, take some amoeba, A, that divides to become two amoebas, B and C. Before the split, both B and C were A. So before the split, $B = C$. After the split, though, B and C are distinct amoebas, $\neg B = C$. So if two things are the same, it does not necessarily follow that they are always going to be the same.

We can't get out of this problem in the same way that we got out of the previous ones. The property of being identical to x at all future times is certainly a property of x. And it doesn't appear to be the case that the property is insufficiently fine-grained. There seems to be no way to make it more precise to avoid the problem.

What else can one say? A natural thought is this. Before the split, B wasn't A: it was only *part of* A. But B is an amoeba, and A is a single-celled creature: it has no parts that are amoebas. So B can't be part of A.

More radically, one might suggest that B and C did not really exist before the split, that they came into existence then. If they did not exist before the split, then they were not A before the split. So it's not the case that $B = C$ before the split. But that seems wrong too. B is not a new amoeba; it is simply A, though some of its properties have changed. If this is not clear, just imagine that C were to die at the split. In this case, we would have no hesitation in saying that B is A. (It would just be like a snake shedding its skin.) Now, the identity of something can't be affected by whether there are *other* things around. So *A* is *B*. Likewise, *A* is *C*.

Of course, one might insist that just because *A* takes on new properties, it is, strictly speaking, a new object; not merely an old object with new properties. So *B* is not really *A*. Likewise *C*. But now we are back with the problem with which we started the chapter.

Main ideas of the chapter

- $m = n$ is true just if the names *m* and *n* refer to the same object.
- If two objects are the same, any property of one is a property of the other (Leibniz's Law).

Chapter 10
Vagueness: how do you stop sliding down a slippery slope?

While we are on the subject of identity, here is another problem about it. Everything wears out in time. Sometimes, parts get replaced. Motor bikes and cars get new clutches; houses get new roofs; and even the individual cells in people's bodies are replaced over time. Changes like this do not affect the identity of the object in question. When I replace the clutch on my bike, it remains the same bike. Now suppose that over a period of a few years, I replace every part of the bike, Black Thunder. Being a careful fellow, I keep all the old parts. When everything has been replaced, I put all the old parts back together to recreate the original bike. But I started off with Black Thunder; and changing one part on a bike does not affect its identity: it is still the same bike. So at each replacement, the machine is still Black Thunder; until, at the end, it is—Black Thunder. But we know that that can't be right. Black Thunder now stands next to it in the garage (Figure 10).

Here is another example of the same problem. A person who is 5 years old is a (biological) child. If someone is a child, they are still a child one second later. In which case, they are still a child one second after that, and one second after that, and one second after that.... So after 630,720,000 seconds, they are still a child. But then they are 25 years old!

10. A bike-rider's dilemma.

Arguments like this are reputed to have been invented by Eubulides, the same Eubulides who invented the liar paradox of Chapter 5. They are now called *sorites paradoxes*. (A standard form of the argument is to the effect that by adding one grain of sand at a time, one can never form a heap; 'sorites' comes from 'soros', the Greek for *heap*.) These are some of the most annoying paradoxes in logic. They arise when the predicate employed ('is Black Thunder', 'is a child') is *vague*, in a certain sense; that is, when its applicability is tolerant with respect to very small changes: if it applies to an object, then a very small change in the object will not alter this fact. Virtually all of the predicates that we employ in normal discourse are vague in this sense: 'is red', 'is awake', 'is happy', 'is drunk'—even 'is dead' (dying takes time). Thus, slippery slope arguments of the sorites kind are potentially endemic in our reasoning.

To focus the issue concerning them, let us look at one of these arguments in more detail. Let Jack be the 5-year-old child. Let a_0 be the sentence 'Jack is a child after 0 seconds'. Let a_1 be the sentence 'Jack is a child after 1 second', and so on. If n is any number, a_n is the sentence 'Jack is a child after n seconds'. Let k be some enormous number, at least as great as 630,720,000. We know that a_0 is true. (After 0 seconds have elapsed, Jack is still 5.)

And for each number, n, we know that $a_n \rightarrow a_{n+1}$. (If Jack is a child at any time, he is a child one second later.) We can chain all these premisses together by a sequence of *modus ponens* inferences, like this:

$$\cfrac{\cfrac{a_0 \quad a_0 \rightarrow a_1}{a_1} \quad a_1 \rightarrow a_2}{a_2}$$

$$\ddots$$

$$\frac{a_{k-1} \quad a_{k-1} \rightarrow a_k}{a_k}$$

The final conclusion is a_k, which we know not to be true. Something has gone wrong, and there doesn't seem much scope to manoeuvre.

So what are we to say? Here is one answer, which is sometimes called *fuzzy logic*. Being a child seems to fade out, gradually, just as being a (biological) adult seems to fade in gradually. It seems natural to suppose that the truth value of 'Jack is a child' also fades from true to false. Truth, then, comes by degrees. Suppose we measure these degrees by numbers between 1 and 0, 1 being complete truth, 0 complete falsity. Every situation, then, assigns each basic sentence such a number.

What about sentences containing operators like negation and conjunction? As Jack gets older, the truth value of 'Jack is a child' goes down. The truth value of 'Jack is not a child' would seem to go up correspondingly. This suggests that the truth value of $\neg a$ is 1 minus the truth value of a. Suppose we write the truth value of a as $|a|$; then we have:

$$|\neg a| = 1 - |a|$$

Here is a table of some sample values:

a	$\neg a$
1	0
0.75	0.25
0.5	0.5
0.25	0.75
0	1

What about the truth value of conjunctions? A conjunction can only be as good as its worst bit. So it's natural to suppose that the truth value of a & b is the *minimum* (lesser) of $|a|$ and $|b|$:

$$|a \& b| = Min(|a|, |b|)$$

Here is a table of some sample values. Values of a are down the left hand column; values of b are along the top row. The corresponding values of a & b are where the appropriate row and column meet. For example, if we want to find $|a \& b|$, where $|a| = 0.25$ and $|b| = 0.5$, we see where the italicized row and column meet. The result is in boldface.

$a \& b$	1	0.75	*0.5*	0.25	0
1	1	0.75	*0.5*	0.25	0
0.75	0.75	0.75	*0.5*	0.25	0
0.5	0.5	0.5	*0.5*	0.25	0
0.25	*0.25*	*0.25*	***0.25***	*0.25*	*0*
0	0	0	*0*	0	0

Similarly, the value of a disjunction is the *maximum* (greater) of the values of the disjuncts:

$$|a \vee b| = Max(|a|,|b|)$$

I leave it to you to construct a table of some sample values. Notice that, according to the above, $\neg$, &, and $\vee$ are still truth functions. That is, for example, the truth value of a & b is determined by the truth values of a and b. It is just that those values are now numbers between 0 and 1, instead of T and F. (It is perhaps worth noting, though, that if we think of 1 as T, and 0 as F, the results where only 1 and 0 are involved are the same as for the truth functions of Chapter 2, as you can check for yourself.)

What of conditionals? We saw in Chapter 7 that there are good reasons to suppose that $\rightarrow$ is not a truth function, but let us set those worries aside for the present. If it is a truth function, which one is it, now that we have to take into account degrees of truth? No answer seems terribly obvious. Here is one (fairly standard) suggestion, which at least seems to give the right *sorts* of results.

$$\text{If } |a| \leq |b|: \quad |a \rightarrow b| = 1$$
$$\text{If } |b| < |a|: \quad |a \rightarrow b| = 1 - (|a| - |b|)$$

($<$ means 'is less than'; $\leq$ means 'is less than or equal to'.) Thus, if the antecedent is less true than the consequent, the conditional is completely true. If the antecedent is more true than the consequent, then the conditional is less than the maximal truth by the difference between their values (i.e. by the amount that truth goes down). Here is a table of some sample values. (Recall that the values of a are down the left-hand column and those of b are along the top row.)

$a \to b$	1	0.75	0.5	0.25	0
1	1	0.75	0.5	0.25	0
0.75	1	1	0.75	0.5	0.25
0.5	1	1	1	0.75	0.5
0.25	1	1	1	1	0.75
0	1	1	1	1	1

What of validity? An inference is valid if the conclusion holds in every situation where the premisses hold. But what is it now for something to hold in a situation? When it is true enough. But how true is true enough? That will just depend on the context. For example, 'is a new bike' is a vague predicate. If you go to a bike dealer who tells you that a certain bike is new, you expect it never to have been used before. That is, you expect 'This is a new bike' to have value 1. Suppose, on the other hand, that you go to a bike rally, and are asked to pick out the new bikes. You will pick out the bikes that are less than a year or so old. In other words, your criterion for what is acceptable as a new bike is more lax. 'This is a new bike' need have value only, say, 0.9 or greater.

So we suppose that there is some level of acceptability, fixed by the context. This will be a number somewhere between 0 and 1—maybe 1 itself in extreme cases. Let us write this number as ε. Then an inference is valid for that context just if the conclusion has a value at least as great as ε in every situation where the premisses all have values at least as great as ε.

Now, how does all this bear on the sorites paradox? Suppose we have a sorites sequence. As above, let a_n be the sentence 'Jack is a child after n seconds'; but to keep things manageable, let us suppose that Jack grows up in four seconds! Then a record of truth values might be:

a_0	a_1	a_2	a_3	a_4
1	0.75	0.5	0.25	0

$a_0 \to a_1$ has value 0.75 (= (1– (1 – 0.75)); so does $a_1 \to a_2$; in fact, every conditional of the form $a_n \to a_{n+1}$ has the value 0.75.

What this tells us about the sorites paradox depends on the level of acceptability, ε, that is in force here. Suppose the context is one that imposes the highest level of acceptability; ε is 1. In this case, *modus ponens* is valid. For suppose that $|a| = 1$ and $|a \to b| = 1$. Since $|a \to b| = 1$, we must have $|a| \le |b|$. It follows that $|b| = 1$. Thus the sorites argument is valid. In this case, though, each conditional premiss, having value 0.75, is unacceptable.

If, on the other hand, we set the level of acceptability lower than 1, then *modus ponens* turns out to be invalid. Suppose, for the sake of illustration, that ε is 0.75. As we have already seen, a_1 and $a_1 \to a_2$ both have value 0.75, but a_2 has value 0.5, which is less than 0.75.

Either way you look at it, then, the argument fails. Either some of the premisses aren't acceptable; or, if they are, the conclusions don't follow validly. Why are we taken in by sorites arguments so easily? Maybe because we confuse complete truth with near-complete truth. A failure to draw the distinction doesn't make much difference normally. But if you do it again, and again, and again... it does.

That's one diagnosis of the problem. But with vagueness, nothing is straightforward. What was the problem about saying that 'Jack is a child' is simply true, until a particular point in time, when it becomes simply false? Just that there seems to be no such point.

Any place one chooses to draw the line is completely arbitrary; it can be, at best, a matter of convention. But now, at what point in Jack's growing up does he cease to be 100 per cent a child; that is, at what point does 'Jack is a child' change from having the value of exactly 1, to a value below 1? Any place one chooses to draw this line would seem to be just as arbitrary as before. (This is sometimes called the problem of *higher-order vagueness*.) If that is right, we haven't really solved the most fundamental problem about vagueness: we have just relocated it.

Main ideas of the chapter

- Truth values are numbers between 0 and 1 (inclusive).
- $|\neg a| = 1 - |a|$
- $|a \vee b| = Max(|a|, |b|)$
- $|a \,\&\, b| = Min(|a|, |b|)$
- $|a \rightarrow b| = 1$ if $|a| \leq |b|$; $|a \rightarrow b| = 1 - (|a| - |b|)$ otherwise.
- A sentence is true in a situation just if its truth value is at least as great as the (contextually determined) level of acceptability.

Chapter 11
Probability: the strange case of the missing reference class

The preceding chapters have given us at least some feel for which inferences are deductively valid, and why. It's now time to come back to the question of inductive validity: that is, the validity of those inferences where the premisses give some ground for the conclusion; yet where, even if the premisses are true in some situation, the conclusion could still turn out to be false.

As I noted in Chapter 1, Sherlock Holmes was very good at this kind of inference. Let us start with an example from him. The mystery of *The Red-Headed League* commences when Holmes and Dr Watson receive a visit from a certain Mr Jabez Wilson. When Wilson enters, Watson looks to see what Holmes has inferred about him (Figure 11):

> 'Beyond the obvious fact that he has at some time done manual labour, that he takes snuff, that he is a Freemason, that he has been in China, and that he has done a considerable amount of writing lately, I can deduce nothing else.'
>
> Mr. Jabez Wilson started up in his chair with his forefinger upon the paper, but his eyes upon my companion.
>
> 'How, in the name of good fortune, did you know all that, Mr. Holmes?' he asked.

11. Holmes displays his logical prowess.

Holmes is pleased to explain. For example, concerning the writing:

> 'What else can be indicated by that right cuff so very shiny for five inches, and the left one with the smooth patch near the elbow where you rest it upon the desk.'

Despite the fact that Holmes is wont to call this kind of inference a deduction, the inference is, in fact, an inductive one. It is entirely possible that Wilson's coat should have shown these patterns without his having done much writing. He could, for example, have stolen it from someone who had. None the less, the inference is clearly a pretty good one. What makes it, and inferences like it, good? One plausible answer is in terms of probability. So let's talk about this, and then we can return to the question.

A probability is a number assigned to a sentence, which measures how likely it is, in some sense, that the sentence is true. Let us write $pr(a)$ for the probability of a. Conventionally, we measure probabilities on a scale between 0 and 1. If $pr(a) = 0$, a is certainly false; then as $pr(a)$ increases, it gets more likely that a is true; until when $pr(a) = 1$, a is certainly true.

What else can one say about these numbers? Let me illustrate with a simple example. Suppose we consider the days of any one particular week. Let w be a sentence that is either true or false every day—say, 'It is warm'—and let r be another—say, 'It is raining'. Let the relevant information be given by the following table:

	Mon	Tue	Wed	Thu	Fri	Sat	Sun
w			✓	✓		✓	✓
r		✓	✓			✓	

A tick indicates that the sentence is true that day; a blank that it is not.

Now, if we are talking about this particular week, what is the probability that on any day, chosen at random, it was warm? There were four warm days, and seven days in total. So the probability is 4/7. Similarly, there were three days where it rained, so the probability that it rained is 3/7:

$$pr(w) = 4/7$$
$$pr(r) = 3/7$$

In general, if we write $\#a$ to mean the number of days at which the sentence a is true, and N for the total number of days:

$$pr(a) = \#a/N$$

How does probability relate to negation, conjunction, and disjunction? Negation first. What is the probability of $\neg w$? Well, there were three days on which it was not warm, so $pr(\neg w) = 3/7$. Notice that $pr(w)$ and $pr(\neg w)$ add up to 1. This is no accident. We have:

$$\#w + \#\neg w = N$$

Dividing both sides by N, we get:

$$\frac{\#w}{N} + \frac{\#\neg w}{N} = 1$$

That is, $pr(w) + pr(\neg w) = 1$.

For conjunction and disjunction: there are two days on which it was both warm and rainy, so $pr(w \,\&\, r) = \#(w \,\&\, r)/N = 2/7$. And there are five days on which it was either warm or rainy, so $pr(w \vee r) = \#(w \vee r)/N = 5/7$. What is the relation between these two numbers? To find the number of days when $w \vee r$ is true, we can start by adding up the days when w is true, then add the number of days where r is true. This won't quite do, since some days will have been counted twice: Wednesday and Saturday. These are the days when it was both rainy and warm. So to get the correct figure, we have to subtract the number of days when it was both:

$$\#(w \vee r) = \#w + \#r - \#(w \,\&\, r)$$

Dividing both sides by N, we get:

$$\frac{\#(w \vee r)}{N} = \frac{\#w}{N} + \frac{\#r}{N} - \frac{\#(w \,\&\, r)}{N}$$

That is:

$$pr(w \vee r) = pr(w) + pr(r) - pr(w \,\&\, r)$$

This is the general relationship between probabilities of conjunctions and of disjunctions.

In Chapter 10, we saw that degrees of truth can also be measured by numbers between 0 and 1, and it might be natural to suppose that degrees of truth and probabilities are the same. They are not. In particular, conjunction and disjunction work quite differently. For degrees of truth, disjunction is a truth function. Specifically, $|w \vee r|$ is the maximum of $|w|$ and $|r|$. But $pr(w \vee r)$ is not determined by $pr(w)$ and $pr(r)$ alone, as we have just seen. In particular, for our w and r, $pr(w) = 4/7, pr(r) = 3/7$, and $pr(w \vee r) = 5/7$. But if $|w| = 4/7$ and $|r| = 3/7$, $|w \vee r| = 4/7$, not 5/7.

Before we can get back to inductive inferences, there is one more bit of information about probability that we need. Given our sample week, the probability that it was raining on some day, chosen at random, is 3/7. But suppose you know that the day in question was a warm one. What is the probability now that it rained? Well, there were four warm days, but only two of those were rainy, so the probability is 2/4. This figure is called a *conditional probability*, and written like this: $pr(r|w)$, the probability of r given w. If we think about it a little, we can give a general formula for calculating conditional probabilities. How did we arrive at the figure 2/4? First, we restricted ourselves to those days when w is true; then we divided this into the number of those days when r is true, that is, the number of days when both w and r are true. In other words:

$$pr(r \mid w) = \#(w \,\&\, r) \div \#w$$

A little algebra tells us that this is equal to:

$$\frac{\#(w \,\&\, r)}{N} \div \frac{\#w}{N}$$

And this is $pr(w \& r) \div pr(w)$.

So here is our general formula for conditional probability:

CP : $pr(r \mid w) = pr(w \& r) / pr(w)$

A modicum of care is required in applying this formula. Dividing by the number 0 makes no sense. 3/0, for example, has no value. Mathematicians call this ratio *undefined*. In the formula for $pr(r|w)$, we have divided by $pr(w)$, which makes sense only if this is not zero, that is, only if w is true at least sometimes. Otherwise, the conditional probability is undefined.

Now, at last, we can come back to inductive inferences. What is it for an inference to be inductively valid? Simply that the premisses make the conclusion more probable than not. That is, the conditional probability of c, the conclusion, given p, the premiss (or the conjunction of the premisses if there are more than one) is greater than that of the negation of c:

$$pr(c \mid p) > pr(\neg c \mid p)$$

Thus, if we are reasoning about the week of our illustration, the inference:

It was a rainy day; so it was a warm day;

is inductively valid. As is easy to check, $pr(w|r) = 2/3$, and $pr(\neg w|r) = 1/3$.

The analysis can be applied to show why the inference of Holmes with which we started is valid. Holmes concluded that Jabez Wilson had been doing a lot of writing (c). His premiss was to the effect that there were certain marks of wear on Wilson's jacket (p). Now, had we gone around the London of Holmes's day, and collected all those people with worn cuffs of the kind in question, then the majority of those would have been clerks, people who spent their working lives writing—or so we may suppose. Thus, the probability that Jabez had been doing a lot of writing, given that his coat bore those marks, is greater than the probability that he had not. Holmes's inference is indeed inductively valid.

Let me finish by noting one puzzle to which the machinery we have just deployed gives rise. As we have seen, a probability can be calculated as a ratio: we take a certain *reference class*; then we calculate the numbers of various groups within it; then we do some dividing. But which reference class do we use? In the illustrative example concerning the weather, I started by specifying the reference class in question: the days of that particular week. But real-life problems are not posed in this way.

Come back to Jabez Wilson. To work out the probabilities relevant in this case, I suggested that we take the reference class to comprise the people living in London in Holmes's day. But why this? Why not the people living in the whole of England then, or in Europe, or just the males in London, or just the people who could afford to come and see Holmes? Maybe, in some of these cases, it wouldn't make much difference. But certainly in others it would. For example, the people who came to see Holmes were all relatively well off, and not likely to wear second-hand coats. Things would be quite different with a wider population. So what should the appropriate reference class have been? This is the sort of question that keeps actuaries (the people who try to figure out risk-factors for insurance companies) awake at night.

In the last analysis, the most accurate reference class would seem to be the one comprising just Wilson himself. After all, what do facts about *other* people ultimately have to do with him? But in that case, he had either been doing a lot of writing, or he had not. In the first case, the probability that he has been writing given that he has a shiny cuff, is 1, and the inference is valid; in the second, it is 0, and the inference is not valid. In other words, the validity of the inference depends entirely on the truth of the conclusion. So you can't employ the inference in order to *determine* whether or not the conclusion is true. If we go this far, the notion of validity delivered is entirely useless.

Main ideas of the chapter

- The probability of a statement is the number of cases in which it is true, divided by the number of cases in the reference class.
- $pr(\neg a)=1-pr(a)$
- $pr(a \vee b)= pr(a)+pr(b)-pr(a \& b)$
- $pr(a \mid b)=pr(a \& b) / pr(b)$
- An inference is inductively valid just if the conditional probability of the conclusion given the (conjunction of the) premiss(es) is greater than that of its negation given the premisses.

Chapter 12
Inverse probability: you can't be indifferent about it!

Chapter 11 gave us a basic understanding of probability and the role it may have in inductive inferences. In this chapter, we'll look at some further aspects of this. Let's start by considering a very famous inductive inference.

The physical cosmos is not a purely random mess. It shows very distinctive patterns: matter is structured into galaxies (Figure 12), which are structured, in turn, into stars and planetary systems, and on some of those planetary systems, matter is structured in such a way as to produce living creatures like you and me. What is the explanation for this? You might say that the explanation is provided by the laws of physics and biology. And so it may be. But why are the laws of physics and biology the way they are? After all, they could have been quite different. For example, gravity could have been a force of *repulsion*, not attraction. In that case, there would never have been stable chunks of matter, and life as we know it would have been impossible anywhere in the cosmos. Does this not give us excellent reason to believe in the existence of a creator of the cosmos: an intelligent being who brought into existence the cosmos, together with its physical and biological laws, for some purpose or other? In short, does not the fact that the physical cosmos is ordered in the way that it is give us reason to believe in the existence of a god of a certain kind?

12. Matter has a distinctive structure. A spiral galaxy.

This argument (for the existence of god) is often called the 'Argument from Design'. It might better be called the Argument *to* Design; but never mind that. Let us think about it more closely. The premiss of the argument, o, is a statement to the effect that the cosmos is ordered in a certain way. The conclusion, g, asserts the existence of a creator-god. Unless g were true, o would be most unlikely; so, the argument goes, given that o, g is likely. Now, it is certainly true that the conditional probability of o given that g is true, is much higher than that of o given that g is false:

1 $pr(o|g) > pr(o|\neg g)$

But this does not give us what we want. For o to be a good inductive reason for g, we need the probability of g, given o, to be greater than that of its negation:

2 $pr(g|o) > pr(\neg g|o)$

And the fact that $pr(o|g)$ is high does not necessarily mean that $pr(g|o)$ is high. For example, the probability that you are in Australia, given that you see a kangaroo in the wild, is very high. (Everywhere else, it would have to have escaped from a zoo.) But the probability that you will see a kangaroo in the wild, given that you are in Australia, is very low. (I lived in Australia for about ten years before I saw one.)

$pr(o|g)$ and $pr(g|o)$ are called *inverse probabilities*, and what we have seen is that for the design argument to work, the relationship between them must be such as to get us from 1 to 2. Is it? There is, in fact, a very simple relationship between inverse probabilities. Recall from the equation **CP** of Chapter 11 that, by definition:

$$pr(a \mid b) = pr(a \,\&\, b) \,/\, pr(b)$$

So:

3 $pr(a|b) \times pr(b) = pr(a \,\&\, b)$

Similarly:

$$pr(b \mid a) = pr(b \,\&\, a) \,/\, pr(a)$$

So:

4 $pr(b|a) \times pr(a) = pr(b \,\&\, a)$.

But $pr(a \,\&\, b) = pr(b \,\&\, a)$ (since a & b and b & a are true in exactly the same situations). Thus, 3 and 4 give us:

$$pr(a \mid b) \times pr(b) = pr(b \mid a) \times pr(a)$$

Assuming that $pr(b)$ is not 0—I shall make assumptions of this kind without further mention—we can rearrange this equation to get:

Inv: $pr(a|b) = pr(b|a) \times pr(a) / pr(b)$

This is the relationship between inverse probabilities. (To remember this, it may help to note that on the right hand side, it's first a b followed by an a, and then an a followed by a b.)

Using **Inv** to rewrite the inverse probabilities in 1, we get:

$$pr(g \mid o) \times \frac{pr(o)}{pr(g)} > pr(\neg g \mid o) \times \frac{pr(o)}{pr(\neg g)}$$

And cancelling the $pr(o)$ on both sides gives:

$$\frac{pr(g \mid o)}{pr(g)} > \frac{pr(\neg g \mid o)}{pr(\neg g)}$$

Or, rearranging the equation:

5 $$\frac{pr(g \mid o)}{pr(\neg g \mid o)} > \frac{pr(g)}{pr(\neg g)}$$

Recall that for the Argument to Design to work, we have to get to 2, which is equivalent to:

$$\frac{pr(g \mid o)}{pr(\neg g \mid o)} > 1$$

It would appear that the only plausible thing that will take us to this from 5 is $\frac{pr(g)}{pr(\neg g)} \geq 1$, that is:

$$pr(g) \geq pr(\neg g)$$

The values $pr(g)$ and $pr(\neg g)$ are called *prior probabilities*; that is, the probabilities of g and $\neg g$ prior to the application of any evidence, such as o. Hence, what we seem to need to make the Argument go through is that the prior probability that there is a creator-god is greater than (or equal to) the prior probability that there is not.

Is it? Unfortunately, there is no reason to believe so. In fact, it would seem that it is the other way around. Suppose you don't know what day of the week it is. Let m be the hypothesis that it is

Monday. Then $\neg m$ is the hypothesis that it is not Monday. Which is more likely, m or $\neg m$? Surely, $\neg m$: because there are lots more ways for it not to be Monday than there are for it to be Monday. (It could be Tuesday, Wednesday, Thursday, ...) Similarly with god. Conceivably, there are many different ways that the cosmos could have been. And intuitively, relatively few of those are significantly ordered: order is something *special*. That, after all, is what gives the Argument to Design its bite. But then there are relatively few possible cosmoses in which there is an *orderer*. So a priori, it is much more likely that there is no creator than that there is.

What we see, then, is that the Argument to Design fails. It is seductive because people often confuse probabilities with their inverses, and so slide over a crucial part of the argument.

Many inductive arguments require us to reason about inverse probabilities. The Argument to Design is not special in this regard. But many arguments are more successful in doing this. Let me illustrate. Suppose you visit the local casino. They have two roulette wheels. Call them A and B. You have been told by a friend that one of them is fixed—though the friend couldn't tell you which one. Instead of coming up red half of the time, and black half of the time, as a fair wheel should, it comes up red 3/4 of the time, and black 1/4 of the time. (Strictly speaking, real roulette wheels come up green occasionally as well; but let's ignore this fact to keep things simple.) Now, suppose you watch one of the wheels, say wheel A, and on five successive spins it comes up with the results:

R, R, R, R, B

(R is red, B is black). Are you justified in inferring that this is the wheel that is fixed? In other words, let c be a statement to the effect that this particular sequence came up, and f be the statement that wheel A is fixed. Is the inference from c to f a good inductive inference?

We need to know whether $pr(f|c) > pr(\neg f|c)$. Using the equation **Inv** to convert this into a relationship between inverse probabilities, what this means is that:

$$pr(c|f) \times \frac{pr(f)}{pr(c)} > pr(c|\neg f) \times \frac{pr(\neg f)}{pr(c)}$$

Multiplying both sides by $pr(c)$ gives:

$$pr(c|f) \times pr(f) > pr(c|\neg f) \times pr(\neg f)$$

Is this true? For a start, what are the prior probabilities of f and $\neg f$? We know that either A or B is fixed (but not both). We have no more reason to believe that it is wheel A, rather than wheel B, or vice versa. So the probability that it is wheel A is 1/2, and the probability that it is wheel B is also 1/2. In other words, $pr(f)=1/2$, and $pr(\neg f) = 1/2$. So we can cancel these out so that the relevant condition becomes:

$$pr(c|f) > pr(c|\neg f)$$

The probability of observing the sequence stated by c, given that the wheel is fixed in the way described, $pr(c|f)$, is $(3/4)^4 \times (1/4)$. (Never mind if you don't know why: you can take my word for it.) This is $81/4^5$, which works out to 0.079. The probability that the sequence is observed, given that the wheel is not fixed, and so fair, $pr(c|\neg f)$, is $(1/2)^5$ (again, take my word for this if you wish), which works out to 0.031. This is less than 0.079. So the inference is valid.

The way that we worked out prior probabilities here is worth noting. We have two possibilities: either wheel A is fixed, or wheel B is. And we have no information that distinguishes between these two possibilities. So we assign them the same probability. This is an application of something called the *Principle of Indifference*. The Principle tells us that when we have a number of possibilities, with no relevant difference between any of them, they all have the same probability. Thus, if there are N possibilities in all, each has probability $1/N$. The Principle of Indifference is a sort of *symmetry* principle.

Notice that we could not apply the Principle in the Argument to Design. In the roulette case, there are two possible situations which are completely symmetric: wheel A is fixed; wheel B is fixed. In the Argument to Design, there are two situations: a creator-god exists; a creator-god does not exist. But these two situations are no more symmetric than: today is Monday; today is not Monday. As we saw, intuitively, there are lots more possibilities in which there is no creator than possibilities in which there is.

The Principle of Indifference is an important part of intuitive reasoning about probability. Let us end this chapter by noting that it is not without its problems. It is well known that it leads to paradoxes in certain applications. Here is one.

Suppose a car leaves Brisbane at noon, travelling to a town 300 km away. The car averages a constant velocity somewhere between 50 km/h and 100 km/h. What can we say about the probability of the time of its arrival? Well, if it is going at 100 km/h it will arrive at 3 p.m.; and if it is going at 50 km/h, it will arrive at 6 p.m. Hence, it will arrive between these two times. The mid-point between these times is 4.30 p.m. So by the Principle of Indifference, the car is as likely to arrive before 4.30 p.m. as after it. But now, half way between 50 km/h and 100 km/h is 75 km/h. So again by the Principle of Indifference, the car is as likely to be travelling over 75 km/h as under 75 km/h. If it is travelling at 75 km/h, it will arrive at 4 p.m. So it is as likely to arrive before 4 p.m. as after it. In particular, then, it is *more* likely to arrive before 4.30 p.m. than after it. (That gives it an *extra* half an hour.)

I'll leave you to think about this. We have had quite enough about probability for one chapter!

Main ideas of the chapter

- $pr(a \mid b) = pr(b \mid a) \times \frac{pr(a)}{pr(b)}$
- Given a number of possibilities, with no relevant difference between them, they all have the same probability (Principle of Indifference).

Chapter 13
Decision theory: great expectations

Let us look at one final issue concerning inductive reasoning. This topic is sometimes called *practical reasoning*, since it is reasoning about how one should act. Here is a famous piece of practical reasoning.

You can choose to believe in the existence of (a Christian) God; you can choose not to. Let us suppose that you choose to believe. Either God exists or God does not. If God exists, all well and good. If not, then your belief is a minor inconvenience: it means that you will have wasted a bit of time in church, and maybe done a few other things that you would not otherwise have wanted to do; but none of this is disastrous. Now suppose, on the other hand, that you choose not to believe in the existence of God. Again, either God exists or not. If God does not exist, all well and good. But if God *does* exist, boy are you in trouble! You are in for a lot of suffering in the afterlife; maybe for all eternity if a bit of mercy isn't thrown in. So any wise person ought to believe in the existence of God. It's the only prudent course of action.

The argument is now usually called *Pascal's Wager*, after the 17th-century philosopher Blaise Pascal who first put it forward. What is one to say about the Wager?

Let us think a little about how this kind of reasoning works, starting with a slightly less contentious example. When we perform actions, we often cannot be sure of the results, which may not be entirely under our control. But we can usually estimate how likely the various possible results are; and, just as importantly, we can estimate the *value* to ourselves of the various results. Conventionally, we can measure the value of an outcome by assigning it a number on the following scale, open ended in both directions:

$$\ldots, -4, -3, -2, 0, +1, +2, +3, +4, \ldots$$

Positive numbers are good, and the further to the right, the better. Negative numbers are bad, and the further to the left, the worse. 0 is a point of indifference: we don't care either way.

Now, suppose there is some action we may perform, say going for a bike ride. It may, however, rain. A bike ride when it is not raining is great fun, so we would value that at, say, +10. But a bike ride when it is raining can be pretty miserable, so we would value that at, say, −5. What value should be put on the only thing that is under our control: going on the ride? We could just add the two figures, −5 and 10, together, but that would be missing an important part of the picture. It may be that it is most unlikely to rain, so although the possibility of rain is bad, we do not want to give it too much weight. Suppose the probability of rain is, say, 0.1; correspondingly, the probability of no rain is 0.9. Then we can weight the values with the appropriate probabilities to arrive at an overall value:

$$0.1 \times (-5) + 0.9 \times 10$$

This is equal to 8.5, and is called the *expectation* of the action in question, going for a ride. ('Expectation', here, is a technical term; it has virtually nothing to do with the meaning of the word as used normally in English.)

In general, let a be the statement that we perform some action or other. Suppose, for simplicity, that there are two possible outcomes; let o_1 state that one of these occurs, and let o_2 state that the other occurs. Finally, let $V(o)$ be the value we attach to o being true. Then the expectation of a, $E(a)$, is the number defined by:

$$pr(o_1) \times V(o_1) + pr(o_2) \times V(o_2)$$

(Strictly speaking, the probabilities in question should be conditional probabilities, $pr(o_1|a)$ and $pr(o_2|a)$, respectively. But in the example, going for a ride has no effect on the probability of rain. The same is true in all the other examples we will look at. So we can stick with the simple prior probabilities here.)

So far so good. But how does this help me to decide whether or not to go for the bike ride? I know the overall value of my going for a ride. Its expectation is 8.5, as we have just seen. What is the expectation of not going for a ride? Again, either it will rain or it will not—with the same probabilities. The two outcomes now are (i) that it will rain and I stay at home; and (ii) that it will not rain and I stay at home. In each of these cases, I derive no pleasure from a bike ride. It might be slightly worse if it doesn't rain. In that case I might be annoyed that I didn't go. But in neither case is it as bad as getting soaked. So the values might be 0 if it rains, and −1 if it does not. I can now calculate the expectation of staying at home:

$$0.1 \times 0 + 0.9 \times (-1)$$

This comes to −0.9, and gives me the information I need; for I should choose that action which has the highest overall value, that is, expectation. In this case, going has expectation 8.5, whilst staying at home has value −0.9. So I should go for a ride.

Thus, given a choice between a and $\neg a$, I should choose whichever has the greater expectation. (If they are the same, I can simply choose at random, say by tossing a coin.) In the previous case, there are only two possibilities. In general, there might be more (say, going for a ride, going to the movies, and staying at home). The principle is the same, though: I calculate the expectation of each possibility, and choose whichever has the greatest expectation. This sort of reasoning is a simple example from the branch of logic called *decision theory*.

Now let's come back to Pascal's Wager. In this case, there are two possible actions: believing or not; and there are two relevant possibilities: God exists or does not. We can represent the relevant information in the form of the following table.

	God exists	God doesn't exist
I believe (b)	$0.1\backslash+10^2$	$0.9\backslash-10$
I don't believe ($\neg b$)	$0.1\backslash-10^6$	$0.9\backslash+10^2$

The figures to the left of the backward slashes are the relevant probabilities, 0.1 that God exists, say, and 0.9 that God doesn't exist. (Whether or not I believe has no effect on whether or not God exists, so the probabilities are the same in both rows.) The figures to the right of the slashes are the relevant values. I don't mind too much whether or not God exists; the important thing is that I get it right; so the value in each of these cases is $+10^2$. (Perhaps one's preferences here might not be exactly the same, but it doesn't matter too much, as we shall see.) Believing, when God doesn't exist, is a minor inconvenience, so gets the value -10. Not believing, when God does exist, is really bad, though. It gets the value -10^6.

Given these values, we can compute the relevant expectations:

$$E(b) = 0.1 \times 10^2 + 0.9 \times (-10) \simeq 0$$
$$E(\neg b) = 0.1 \times (-10^6) + 0.9 \times 10^2 \simeq -10^5$$

($\simeq$ means 'is approximately equal to'.) I should choose whichever action has the greater expectation, which is to believe.

You may think that the precise values I have chosen are somewhat artificial; and so they are. But in fact, the precise values don't really matter too much. The important one is the -10^6. This figure represents something that is really bad. (Sometimes, a decision theorist might write this as $-\infty$.) It is so bad that it will swamp all the other figures, even if the probability of God's existence is very low. That is the punch in Pascal's Wager.

The Wager might look fairly persuasive, but in fact it makes a rather simple decision-theoretic error. It omits some relevant possibilities. There is not just one possible god, there are many: a Christian god (God), Islam's Allah, Hinduism's Brahman, and lots more that various minor religions worship. And a number of these are very jealous gods. If God exists, and you don't believe, you are in trouble; but if Allah exists and you don't believe, you are equally in trouble; and so on. Moreover, if God exists, and you believe in Allah—or vice versa—this is even worse. For in both Christianity and Islam, believing in false gods is worse than being a simple non-believer.

Let's draw up a table with some more realistic information.

	No God exists	God exists	Allah exists	...
No belief (n)	0.9\+10^2	0.01\−10^6	0.01\−10^6	...
Believe in God (g)	0.9\−10	0.01\+10^2	0.01\−10^9	...
Believe in Allah (a)	0.9\−10	0.01\−10^9	0.01\+10^2	...
⋮	⋮	⋮	⋮	

If we compute the expectations on even this limited amount of information, we get:

$$E(n) = 0.9 \times 10^2 + 0.01 \times (-10^6) + 0.01 \times (-10^6) \approx -2 \times 10^4$$
$$E(g) = 0.9 \times (-10) + 0.01 \times 10^2 + 0.01 \times (-10^9) \approx -10^7$$
$$E(a) = 0.9 \times (-10) + 0.01 \times (-10^9) + 0.01 \times 10^2 \approx -10^7$$

Things are looking pretty bleak all round. But it is clear that theistic beliefs are coming off worst. You shouldn't have any of them.

Let me end, as I have ended all the chapters, with some reasons as to why one might be worried about the general framework deployed—specifically, in this case, the policy of deciding according to the greatest expectation. There are situations where this definitely seems to give the wrong results.

Let's suppose you take the wrong gamble on Pascal's Wager, and end up in Hell. After a few days, the Devil appears with an offer. God has commanded that you be shown some mercy. So the Devil has hatched a plan (Figure 13). He will give you one chance to get out of Hell. You can toss a coin; if it comes down heads, you are out and go to Heaven. If it comes down tails, you stay in Hell forever. The coin is not a fair one, however, and the Devil has control of the odds. If you toss it today, the chance of heads is 1/2 (i.e. 1 – 1/2). If you wait till tomorrow, the chances go up to 3/4 (i.e. $1 - 1/2^2$). You sum up the information:

	Escape	Stay in Hell
Toss today (d)	$0.5\backslash{+10^6}$	$0.5\backslash{-10^6}$
Toss tomorrow (m)	$0.75\backslash{+10^6}$	$0.25\backslash{-10^6}$

Escaping has a very large positive value; staying in Hell has a very large negative value. Moreover, these values are the same today as

13. A devilish plan: never do today what you should put off until tomorrow.

tomorrow. It is true that if you wait till tomorrow, you might have to spend an *extra* day in Hell, but one day is negligible compared with the infinite number of days that are to follow. Then you do the calculations:

$$E(d) = 0.5 \times 10^6 + 0.5 \times (-10^6) = 0$$
$$E(m) = 0.75 \times 10^6 + 0.25 \times (-10^6) = 0.5 \times 10^6$$

So you decide to wait till tomorrow.

But tomorrow the Devil comes to you and says that if you wait one more day, the odds will get even better: they will go up to 7/8 (i.e. $1 - 1/2^3$). I will let you do the calculations: you should decide to wait till the next day. The trouble is that *every* day the Devil comes to you and offers you better odds if you will wait till the next day. The odds get better, day by day, as follows:

$$1-1/2, 1-1/2^2, 1-1/2^3, 1-1/2^4, \ldots, 1-1/2^n, \ldots$$

Every day you do the calculation. The expectation of tossing on the nth day is:

$$(1-1/2^n)\times 10^6 + 1/2^n \times (-10^6)$$

A little arithmetic tells us that this is $10^6 \times (1-2/2^n) = 10^6 \times (1-1/2^{n-1})$. The expectation for waiting till the next, $n+1$st, day is the same, with n replaced by $n+1$. That is, $10^6 \times (1-1/2^n)$—which is larger. ($1/2^n$ is smaller than $1/2^{n-1}$.) Every day, the expectation goes up.

Hence, every day you do the rational thing and wait till the next day. The result is that you never toss the coin at all, so you stay in Hell for ever! Tossing on *any* day has to be better than that. So it looks as though the only rational thing to do is to be irrational!

Main ideas of the chapter

- $E(a) = pr(o_1)\times V(o_1) + \ldots + pr(o_n)\times V(o_n)$, where $o_1, \ldots, o_n$ state all the possible outcomes that might result from a being true.
- The rational action is the one with the greatest expectation.

Chapter 14
Halt! What goes there?

If you have got this far in the book, you will have a reasonable sense of the basic ideas of modern logic. But that is just a beginning. Modern logic goes a long way beyond these ideas, with results of great profundity and beauty. There is, of course, no possibility of surveying these in a book of this nature, but this chapter and the next will, at least, offer a glimpse of what lies beyond. We will take a look at some results about what formal reasoning can and can't do, and some of the philosophical implications of these facts. A warning: these chapters may be a bit tougher than the previous ones. I have made matters as simple as I can, but we are dealing with some complex mathematical issues. Having said all this, on with the topic of this chapter.

Leibniz—the same Leibniz we met in Chapters 6 and 9—had a dream, a dream that would end disputes. Whenever we have a claim over which there is disputation, we may write it in a suitable language, the *characteristica universalis*. Then, to determine the truth of the claim, *calculemus*—we simply calculate. The language is such that there is a process of computation, a *calculus ratiocinator*, which can be applied to determine whether or not the claim is true.

Though Leibniz did suggest steps in the direction of achieving this project, it was never more than a dream. The mathematics of his day was just not up to tackling the project he envisaged.

The mathematics of our day is. The symbolic languages we have been looking at in previous chapters are such that claims whose truth value is not known (at least, a good many such claims) can be expressed in them. The question that then remains is whether there is an appropriate calculating procedure.

The answer (sadly) is that there is not—even for the very limited domain of mathematical claims. This was proved by the British mathematician Alan Turing (1912–54) in 1936 (Figure 14). Turing is one of the founding figures of modern computer science. Of course, in his day there was nothing like the modern computer, now familiar to most people. But the *theory* of such machines was, in fact, worked out by Turing and others well before such computers existed, leaving others to find out how the ideas could be realized in practice—though Turing himself made some notable advances in the more practical aspects of constructing computing machines; for example, with his work on the Enigma Project, aimed at decoding German naval radio transmissions in the Second World War. As might be expected, the connection between Turing's interest in computation and Leibniz's dream is no coincidence.

What is a computer? At its simplest, it is some device that receives an input or inputs, performs some procedure—mathematicians call it an *algorithm*, the name coming from the Persian mathematician Al Khwārizmī (780–850)—and then (if you are lucky) gives you an output.

The inputs and outputs of modern computers are of different kinds: numbers, text, pictures, sounds. But to the machine, these are all just numbers. That is all it can operate on. The input

14. Alan Turing (1912–54), a founder of modern computer science.

devices of the computer translate the input into a sequence of numbers on which the algorithm operates. The output device reverses the procedure.

The form in which the numbers used by the computer are stored is not the one that is familiar from primary-school arithmetic, though. The storage cells of a computer can be in only one of two states: on or off. So the computer has only two basic bits of information it can employ. One can think of these as 1 and 0. Any number has to be expressed using just these two digits. This is done using binary arithmetic. (That is, how you'd count if you had only two fingers.) In standard (decimal) arithmetic, a numeral is really a way of expressing a sum of powers of 10. Thus, 4,302 is:

$$4 \times 10^3 + 3 \times 10^2 + 0 \times 10^1 + 2 \times 10^0$$

(10^0—indeed, anything raised to the power 0—is just 1). Similarly, a binary numeral represents a sum of powers, but this time powers of 2. So 1,011 is:

$$1 \times 2^3 + 0 \times 2^2 + 1 \times 2^1 + 1 \times 2^0$$

The following table gives the conversion between the first few decimal numerals and the first few binary numerals.

Decimal	Binary
0	0
1	1
2	10
3	11
4	100
5	101

We may therefore take our computation (algorithm) to be something that operates on numbers expressed in this binary way.

So much for the input and output, but what is a computation? A computation is specified by a set of rules of the kind found in standard computer programs. Such programs are written in many different languages, whose precise details are not relevant here. A rather dull program might look something like this:

1. if $x = 0$ output x; else go to line 2
2. let $x := x - 1$
3. go to line 1

The left hand numbers are line numbers. The input is some number, x. The first line tests to see whether this is 0, and, if it is, outputs it. Otherwise it goes to the next line. This reduces x

by 1, and then the computation goes back to line 1. As a little thought shows, what this program does is to take any input, and then go round in a loop, subtracting 1 from it till it is 0, which it then outputs.

So far so good. Next—and this is a really clever thing about modern computers—the computer does not have to wait for someone to input each line of the program as the computation proceeds. The program itself is stored in the computer. Of course, it is stored as a number. The computer has no way of storing anything else. (Indeed, one can think of the *whole state* of the computer at any time as simply one huge string of 1s and 0s—an enormous binary numeral!) One may take the number representing the program and stored in the computer as the 'code number' of the program. If n is any number, let P_n be the program with that code number. (If, because of the way that coding is implemented in the computer, n happens not to be the code number of any program, we can just let P_n be the simple program above, by default.) Strictly speaking, a program itself does not really care how many inputs the algorithm it is running is supposed to have. It just helps itself to any information which resides in the computer when it is told to. By convention, though, we can assume that all the input bits of information are set to 0, except the relevant ones, which are appropriately filled.

Now, sometimes a program with a given input will deliver an output; but sometimes it will just keep on going for ever. So consider the following program:

1. let $x := x + 1$
2. if $x = 0$ output x; else go to line 1

The program takes some input and adds 1 to it. It then tests to see whether it is 0, and, if it is, it outputs x. But of course it is not zero (our binary numbers are always greater than or equal to zero), so we go back to line 1 and repeat the process. We never get to 0 by adding 1, and so the computation never ceases, and just goes on forever, in an endless loop (in practice, until the machine wears out, or x become too big for it to handle). Let us call this program, for future reference, L (for looping).

Well-constructed programs are designed so that this can never happen. The programmer analyses the program to see that it can never go off into an infinite loop of this kind. But can this always be done? Is there an algorithm we can apply to a program (or, more precisely, its code number) and inputs, to determine whether or not a computation with that program and those inputs terminates?

The answer is *no*. And this is what Turing proved. The proof is a relatively simple, but very clever, one. It is by *reductio ad absurdum*. In such a proof, we assume the opposite of what we wish to prove, and then show that this leads to something unacceptable.

So suppose that there were an algorithm which did what is needed. Call this A. Thus, when A is applied to two inputs, n and i, it outputs 1 or 0. 1 means that a computation of the program P_n with input i terminates; 0 means that it does not.

Now consider the following algorithm. Let us call it T (for Turing):

- Run the algorithm A with the inputs x and x. That algorithm terminates giving either a 1 or a 0.
- If it is 0, output 1
- If it is 1, run L with input x.

What does this program with input x do? Essentially, it applies A to determine whether a computation of P_x with input x stops. If it does not, it outputs a 1. In particular, it stops. But if that computation does stop, the whole computation goes off into an infinite loop and never stops.

I have described the program *T* in fairly 'high level' terms. But there is nothing particularly problematic about it. Any skilled programmer who understands how information is coded into the computer and stored in it, and who is using a language that has direct access to this data, can write such a program.

Now, and to complete the proof: *T*, itself, has a code number. Call this *t*. We may run *T* with *t* itself as input. If that computation *stops*, then running *A* with inputs *t* and *t* stops and outputs 1. But then the computation with *T* goes off into an infinite loop and never stops. If, on the other hand, running *T* with input *t does not stop*, then running *A* with inputs *t* and *t* stops and outputs 0. So the computation with *T* stops and outputs 1. Hence, if the computation does not stop, it does stop! Either way, then, we have something impossible. So our original assumption that there is an algorithm *A*, must have been false.

The clever thing about Turing's proof is a certain kind of self-reference. (We met self-reference in Chapter 5.) It takes some putative program and applies it to its very own code. This is sometimes called *diagonalization*, a technique invented by the great German mathematician, Georg Cantor (1845–1918), in his investigations of the infinite. You can see why it is called this by considering the following table:

	0	1	2	3	4	...
0	$\mathbf{a}_{00}$	a_{01}	a_{02}	a_{03}	a_{04}	...
1	a_{10}	$\mathbf{a}_{11}$	a_{12}	a_{13}	a_{14}	...
2	a_{20}	a_{21}	$\mathbf{a}_{22}$	a_{23}	a_{24}	...
3	a_{30}	a_{31}	a_{32}	$\mathbf{a}_{33}$	a_{34}	...
4	a_{40}	a_{41}	a_{42}	a_{43}	$\mathbf{a}_{44}$	...
⋮	⋮	⋮	⋮	⋮	⋮	

Down the left-hand side are the codes of programs. Across the top are inputs. The entry a_{xy} is the output for program (with code) x, when run with input y. If that computation does not terminate, we might indicate this by the symbol ∞. What the algorithm A does—were it to exist—is to compute whether the value of a_{xy} is ∞ or not. T takes the result of that computation on the diagonal (bolded), and operates on it to ensure that T behaves differently from P_x on input x. So T cannot be on the left-hand list. But every program occurs on this list. So T does not exist. And since T was defined unproblematically from the algorithm A, A cannot exist either.

The result is known as the *Halting Theorem*. And what it shows is that there is no algorithm which can determine whether any given program, with any given input, halts (though, of course, we may be able to do this in special cases). And—to return finally to Leibniz's dream—what we see is that there are mathematical questions, such as this one, for which there is no algorithm to determine their truth. Leibniz's dream cannot be realized.

I have ended previous chapters by pointing out why the line run in the chapter might be contestable. Let me end this chapter in the same way. Given standard assumptions of number theory, it is impossible to contest Turing's proof. This is as good a piece of mathematics as it is possible for there to be. But in the argument I have given, there is an assumption which I have made so far without comment. The assumption is that everything which we can recognize as an algorithm can be programmed on a computer. If this is not the case, then Turing's proof shows only that there is no computer program which can determine whether any computation stops. But maybe there could be some other sort of algorithm—maybe one which could be deployed in Leibniz's project.

The claim that one can write a computer program for every algorithm is called the *Church–Turing Thesis*, after Turing and the American mathematician Alonzo Church (1903–95). It is not itself susceptible to mathematical proof, since proof can work only on precisely defined notions; and while the notion of what a computer can do *can* be defined in precise mathematical terms, the notion of an algorithm is merely an informal and intuitive one. An algorithm is, roughly, a procedure that can be done in steps where there is no guessing and no creativity—and those are somewhat vague notions.

The Church–Turing Thesis has for a long time been well accepted by mathematicians. There is a history of attempts to refute it. These all attempted to produce something that can be recognized, intuititively, as algorithmic, but which cannot be programmed on a computer—a matter which *is* susceptible of precise mathematical proof. Such attempts all failed; hence the orthodoxy of the Church–Turing Thesis.

However, there are now areas of research into methods of computation other than the sort employed by a desktop computer. These are sometimes called *hypercomputation*. One example: some methods involved use analogue quantities, as opposed to digital. (Analogue quantities are continuous, like length; whereas digital quantities are discrete, like binary numbers.) Another example: some methods involved appeal to properties of space and time in the General Theory of Relativity, where time can 'speed up'. It is, as yet, too early to be clear what the upshot of such investigations will be.

Main ideas of the chapter

- Algorithms can be assigned code numbers.
- If there were an algorithm, *A*, to determine whether the algorithm with code number x (that is, P_x) terminated when

run with input *y*, we could use this to define an algorithm, *T*, which computes the value of *A* with inputs x and x, and uses the result to ensure that its own output is different from each P_x 'along the diagonal'.

- *T* must itself have a code number, *t*. Running *T* with input *t* would then produce an impossible result.
- Hence there is no such algorithm as *T*, and so no such algorithm as *A*.

Chapter 15
Maybe it is true—but you can't prove it!

The title of this chapter sounds rather like something that a petty criminal might say to a police officer. But in fact, it announces another of the most significant results in logic in the last 100 years. (A quick warning: in previous chapters I have used lower case letters (a, b, etc.) for sentences. In this chapter, I will use upper case letters (A, B, etc.), so as to avoid any possible confusion with numbers.)

Leibniz was not the only logician in the history of the subject to have an ambitious project. Another was one of the most significant mathematicians of the 20th century, David Hilbert (1862–1943), who lived and worked in Göttingen. His project, usually called the *Hilbert Program in the Foundations of Mathematics*, was to prove that mathematics was consistent; that is, to prove that in mathematics one can never prove any things of the form A and $\neg A$. In the 1920s mathematics was still in a state of shock from the discovery of Russell's paradox (which we met in Chapter 5) and others of its ilk: paradoxical arguments which struck at the very core of mathematics. Hilbert wanted to make sure that this was not going to happen again.

One has to be a bit careful here. The proof of consistency would, of course, be a mathematical one. And if mathematics *is* inconsistent, maybe it could prove its *consistency* anyway.

Indeed, if the logic employed is the one we looked at in Chapter 2, if mathematics is inconsistent, it can prove everything! As we saw there, in that logic, everything follows from a contradiction. So the proof of consistency had to be done with a particularly secure kind of mathematical reasoning. Hilbert called this *finitary*; but that is not the relevant part of the story here.

The relevant part is this. To prove something consistent you first have to have a fix on it. And if you are going to apply mathematics to it, you need a precise specification. So, as a preliminary step to his main project, Hilbert required an appropriate axiom system for mathematics, which he could then prove to be consistent.

An axiom system comprises a bunch of *axioms*. These are things that we may accept without proof. (The set of axioms may be finite or infinite. But if it is infinite, we need to be able to tell that something *is* an axiom. Specifically, there should be some algorithm which determines the matter.) A *proof* in the system is just a sequence of statements, each of which is either an axiom or can be deduced from earlier statements in the sequence. The *theorems* of the system are things which occur at the ends of proofs. The theorems, then, are the things that can be deduced, ultimately, from the axioms.

The method of axiomatization is a venerable one in mathematics. It was applied to geometry by the Ancient Greek mathematician Euclid (mid-4th to mid-3rd century BCE). However, perhaps surprisingly, the method was not widely applied in mathematics until the 20th century. Until then, the only parts of mathematics which had been formulated axiomatically were geometry (or more precisely, geometries; by the beginning of the 19th century, mathematicians knew there to be geometries other than Euclid's—non-Euclidean geometries), and some parts of abstract algebra.

Hilbert's ambitious proposal required all of mathematics to be axiomatized. That is, an axiom system was needed, whose theorems were all and only the mathematical claims that are true (however one understands what it is to be true in mathematics). The existence of such an axiom-system was disproved by perhaps the most famous logician of the 20th century, the Austrian mathematician Kurt Gödel (1906–78; Figure 15). What Gödel showed was that such an axiom system cannot be provided even for the fragment of mathematics which concerns natural numbers (0, 1, 2,...), let alone the rest of it. This fragment of mathematics is now usually called *arithmetic*. So what Gödel showed was that, though there may be axiom systems capturing *some* of the truths of arithmetic, there is no axiom system capturing *all* of them. As logicians say, the axiom system must be *incomplete*.

Given the things we learned in Chapter 14, Gödel's result can be shown fairly straightforwardly. The proof is, again, by *reductio ad absurdum*. Take a symbolic language which can talk about numbers, and which has the resources to express the things one can do with them: add, multiply, etc. It is not difficult to produce such a thing. The statement that a computation with program (with code) n and input i terminates can be expressed by a statement in this language. (It takes a bit or work to show this, but it is not difficult.) Call this statement S_{ni}. Now suppose that there *were* a complete axiomatization of arithmetic, that is, an axiom system whose theorems were all and only the truths of this language. Then there is an algorithm for deciding whether S_{ni} is true. We simply start to prove the theorems in a systematic way, ensuring that everything that can be proved is proved sooner or later. (It is not difficult to design such a procedure.) Sooner or later, then, either a proof of S_{ni} or $\neg S_{ni}$ will turn up, deciding the question. (We may not know how long this will take, but that does not matter.) But the Halting Theorem of Chapter 14 tells us that there can be no algorithm for deciding the matter. Hence, there can be no such axiomatization.

15. Kurt Gödel (1906–78), arguably the most famous 20th-century logician.

The proof I have just outlined was not Gödel's proof. Indeed, his proof appeared in 1931, five years before Turing's proof of the Halting Theorem. But Gödel's proof was just as ingenious as Turing's—if not more so—and also deployed a certain kind of self-reference. In outline, it goes as follows.

Suppose that we have an axiom system for arithmetic in a language with sufficient expressive capability. All of its theorems are true, but it may not be able to prove *all* the true statements of the language; so it may not be complete. A mathematical statement of this language is just a piece of text, and as we noted in Chapter 14, such a statement can be coded by a number. A computer program is sequence of statements, and, as we noted in Chapter 14, this, too, can be coded as a number. But a mathematical proof is also just a sequence of statements, and so can be coded as a number in the same way.

Now, consider the statement: x is (the code of) a proof of the statement (with code) y. This itself is a statement about numbers, and it can be expressed by a sentence in the language, $Prov(x, y)$. Moreover, provided the axioms of the system are strong enough:

- if m is indeed (the code of) a proof a theorem (with code) n, $Prov(m, n)$ can be proved in the axiom system

To show these things is actually quite hard, and requires some substantial mathematics. But it can be done, as Gödel showed.

To say that statement (with code) y is provable is just to say that there is a proof of it: $\exists x Prov(x, y)$. But—and here is the really clever part of Gödel's proof—by an ingenious construction, one can find a sentence essentially of the form $\neg\exists x Prov(x, n)$, the code number of which is n itself! In effect, the statement says: *this very statement is not provable (in the system)*! Call this sentence G (for Gödel).

Now, suppose that G, that is, $\neg\exists x Prov(x, n)$, were provable in the axiom system. Then some number, m, would be the code of a proof of G. So $Prov(m, n)$ would be true, and so provable in the axiom system (by the bullet point above). But $\exists x Prov(x, n)$ follows from this, and so is provable. The axiom system, then, is inconsistent. Assuming this not to be the case, then G cannot be proved. But in that case it is true, because that is what it says! Hence there are truths in the language which cannot be proved in the system. So there is no complete axiom system for arithmetic.

Gödel's theorem—however it is proved—shows a clear limitation of the axiomatic method in mathematics (which is not to say that it should not be used: indeed, axiomatization is more of a staple of mathematical methodology nowadays than ever). In particular, it dealt a death blow to Hilbert's Program. Arithmetic cannot be axiomatized—much less the whole of mathematics. Gödel's result has been held to have many other philosophical consequences, concerning the nature of numbers, our knowledge of them, and even the nature of the human mind. The debates still continue, and this is not the place to go into them.

Let me finish, once more, by raising an issue about the Theorem. The mathematics of both proofs I have given is ungainsayable. But both proofs make a certain assumption (which Hilbert certainly took—or at least hoped!—to be the case): that the truths about numbers, and so an axiom system which captures any part of these, is consistent. This is quite explicit in Gödel's proof, but is also present in the first proof I gave. That proof assumes that *exactly one* of S_{ni} and $\neg S_{ni}$ will turn up, settling the matter one way or the other. But if the system is inconsistent, this may well not be the case: both may turn up, leaving the matter moot—to say the least!

Now, as may well be clear to readers, the sentence G employed in Gödel's proof is a very close cousin of the liar paradox, which we met in Chapter 5. Both of them say of a certain sentence that it, itself, does not have some crucial property. Indeed, there is a paradox about provability closely related to the liar paradox. Consider the sentence: *this sentence is not provable.* Suppose that it is provable. Then it is true. So it is not provable. Hence (by *reductio ad absurdum*) it cannot be provable. But we have just proved this, so it *is* provable!

If one tries to carry out reasoning of this kind concerning the sentence G in its axiom system, it cannot be reproduced in the axiom system. Perhaps rather surprisingly, if the system is consistent, the claim that if something is provable in the system it is true, cannot be proved in the system. (This was proved by the German mathematician Martin Löb (1921–2006) in 1955, and so is usually called *Löb's Theorem*.) So the paradox cannot be used to establish the inconsistency of the axiom system. However, it does show that paradoxes of self-reference lurk in the vicinity of arithmetic. Given that, perhaps one should not be quite so confident that the truth about numbers is consistent.

Main ideas of the chapter

- An axiom system for arithmetic is complete if it can prove every true sentence in its language.
- No axiom system in a language with sufficiently expressive power is complete.
- This can be inferred from the Halting Theorem.
- It can also be proved by considering a sentence which says, in effect, *this very sentence is not provable* (*in the system*).

Coda on Gödel's *other* incompleteness theorem

The result we have been looking at in this chapter is sometimes called Gödel's *first* incompleteness theorem. At the same time he proved this, Gödel proved another result, called his *second* incompleteness theorem. This shows, essentially, that if we have a consistent axiom system of the kind we have been looking at, it cannot prove a sentence naturally taken as expressing the consistency of the system—at least if the system is based on the logic we looked at in Chapter 2. In fact, it is quite easy to show this, given Löb's Theorem.

First of all, note that since the system can prove that $\neg 0 = 1$, if it can prove that $0 = 1$ it is inconsistent. Conversely, if it is inconsistent, it can prove that $0 = 1$, because of the inference from a contradiction to any conclusion, which we noted at in Chapter 2. Hence, a simple way to say that the system is consistent is to say that it cannot prove that $0 = 1$.

Now, if A is any sentence of the language of the theory, let us write $\langle A \rangle$ for its code number. Then a slightly more precise statement of Löb's Theorem is this:

- If a system can prove the sentence $\exists x Prov(x, \langle A \rangle) \supset A$, it can prove the sentence A.

That is:

- If a system cannot prove the sentence A, it cannot prove the sentence $\exists x Prov(x, \langle A \rangle) \supset A$

So, as a special case, if it cannot prove 0 = 1, it cannot prove that:

- $\exists x Prov(x, \langle 0 = 1 \rangle) \supset 0 = 1$

In other words, if it is consistent, it cannot prove that $\exists x Prov(x, \langle 0 = 1 \rangle) \supset 0 = 1$.

But then it cannot prove that $\neg \exists x Prov(x, \langle 0 = 1 \rangle)$ either, because, whatever C is, $A \supset C$ follows from $\neg A$ (as the truth table for $\supset$ in Chapter 7 shows: if A is false, $A \supset C$ is true). So if the system is consistent, it cannot prove that $\neg \exists x Prov(x, \langle 0 = 1 \rangle)$, i.e., that it *is* consistent. This is Gödel's second incompleteness theorem.

This theorem dealt a second blow to Hilbert's Program. The aim of the program was, recall, first, to axiomatize mathematics; then, secondly, to prove the axiom system consistent. The proof would of course be a mathematical proof, and so be performable in the system itself. The first incompleteness theorem showed that the first step of the program could not be realized. The second incompleteness theorem showed that the second step could not be.

A little history and some further reading

The ideas that we have been looking at in this book were developed at various different times and places. In this section, I will describe the history of logic, and locate the ideas in their historical context. I will first outline briefly the history of logic in general; then I will go through, chapter by chapter, and explain how the details fit into the bigger picture.

As we go along, I will also give some further reading, where you can follow up a number of the issues if you wish. This is not as easy as might be thought. By and large, logicians, philosophers, and mathematicians prefer to write for each other. Finding things written for relative beginners is not easy, but I have done my best.

In Western intellectual history, there have been three great periods of development in logic, with somewhat barren periods sandwiched between them. The first great period was ancient Greece between about 400 BCE and 200 BCE. The most influential figure here is Aristotle (384–322), whom we met in Chapter 6. Aristotle developed a systematic theory of inferences called 'syllogisms', which have the form:

> All [some] *A*s are [are not] *B*s.
> All [some] *B*s are [are not] *C*s.
> So, all [some] *A*s are [are not] *C*s.

Aristotle lived in Athens much of his life, founded a school of philosophy called the Lyceum, and is usually reckoned to be the founder of Western logic. But at about the same time, there was another flourishing school of logic in Megara, about 50 km west of Athens. Less is known about the Megarian logicians, but they seem to have been particularly interested in conditionals, and also in logical paradoxes. Eubulides (whom we met in Chapters 5 and 10) was a Megarian. Another important philosophical movement started in Athens around 300 BCE. It was called Stoicism, after the porch (Greek, 'stoa') where early meetings were held. Though the philosophical concerns of Stoicism were much wider than logic, logic was an important one of them. It is generally supposed that Megarian logic exerted an influence on the Stoic logicians. At any rate, a major concern of Stoic logicians was the investigation of the behaviour of negation, conjunction, disjunction, and the conditional.

It should also be mentioned that at around the same time as all this was happening in Greece, theories of logic were being developed in India. Important as these theories are, though, they never developed to the sophisticated levels to which logic developed in the West.

The second growth period in Western logic starts in the early middle ages, with Arabic philosophers such as Ibn Rushd (Averroes, 1126–98), but blossoms in the medieval European universities, such as Paris and Oxford, from the 12th to the 14th centuries. The medieval logicians included such notables as Duns Scotus (1266–1308) and William of Ockham (1285–1349), and they systematized and greatly developed the logic that they inherited from ancient Greece. After this period, logic largely stagnated till the second half of the 19th century, the only bright spot on the horizon during this period being Leibniz (1646–1716), whom we met in Chapters 6, 9, and 14. Leibniz anticipated some of the modern developments in logic, but the mathematics of his day was just not up to allowing his ideas to take off.

The development of abstract algebra in the 19th century provided just what was required, and triggered the start of the third, and possibly the greatest, of the three periods. Radically new logical ideas were developed by thinkers such as Frege (1848–1925) and Russell (1872–1970), whom we met in Chapters 2 and 4, respectively. The logical theories developing from this work are normally referred to as *modern logic*, as opposed to the *traditional logic* that preceded it.

Developments in logic continued apace throughout the 20th century, and show no sign of slowing down yet.

A standard history of logic is Kneale and Kneale (1975). This is a little dated now, and is characterized by more optimism than is perhaps justified, in its attitude that early modern logicians had finally got everything pretty much right; but it is still an excellent reference work. Zalta (1995–) can be consulted for many authoritative articles on the topics covered in this book, though some of them tend to be quite technical.

* * *

Chapter 1: Validity

The distinction between deductive and inductive validity goes back to Aristotle. Theories of deductive validity have been articulated since that time. The view described in Chapter 1—that an inference is deductively valid just if the conclusion is true in any situation where its premisses are true—can be traced back to medieval logic; but its articulation is a central part of modern logic. A warning: what I have called a *situation* is more commonly called an *interpretation*, *structure*, or sometimes, *model*. The word 'situation' itself has a different, and technical, sense in one area of logic. Lewis Carroll (whose real name was Charles Dodgson) was no mean logician himself, and published a number of works on traditional logic.

Chapter 2: Truth functions—or not?

The argument to the effect that contradictions imply everything is a medieval invention. Exactly who invented it is unclear, but it is certainly to be found in Scotus. The truth-functional understanding of negation, conjunction, and disjunction itself seems to have arisen in the Middle Ages. (The Stoic account was not truth-functional in the modern sense.) In its fully articulated form, it appears in the founders of modern logic, Frege and Russell. A modern dissident is Strawson (1952, ch. 3).

Chapter 3: Names and quantifiers

The distinction between names and quantifiers is largely a creature of modern logic. Indeed, the analysis of quantifiers is often reckoned to be a defining moment in modern logic. It was provided by Frege, and later taken up by Russell. At around the same time, the US philosopher and logician, C. S. Peirce, was developing similar ideas. $\exists x$ is often called the *existential* quantifier; but this terminology smuggles in a somewhat contentious theory of existence. Lewis Carroll's works on *Alice* are replete with philosophical jokes. For an excellent commentary on them, see Heath (1974). For many of Heath's own jokes about nothing, see Heath (1967).

The theories explained in Chapters 1–3 can be found in any standard modern logic text. Hodges (1977) is one that is not pitched at too formidable a level; neither is Lemmon (1971). The techniques of deductive logic to be found in subsequent chapters are described at much greater length in Priest (2008), though this presupposes a first course in logic.

Chapter 4: Descriptions and existence

The isolation of descriptions as an important logical category is also something to be found only in modern logic. Perhaps the most famous analysis of them was given by Russell in 1905. The account given in this chapter is not Russell's, but it is very close in spirit. Descriptions are discussed in some, but not all, standard modern logic texts. Hodges (1977) has a good clear account.

Chapter 5: Self-reference

Various different versions of the liar paradox can be found in ancient Greek philosophy. More paradoxes of self-reference were invented and discussed throughout medieval logic. Even more were discovered around the turn of the 20th century—and this time at the very core of mathematics itself. Since then, they have become a very central issue in logic. Suggestions for solving them are legion. The idea that there might be some sentences that are neither true nor false goes back to Aristotle (*De Interpretatione*, ch. 9); however, he would have had no sympathy with the symmetric idea that some sentences might be both true and false. That there might be such sentences, and that paradoxical sentences might be amongst them, is an unorthodox view that has been advanced by some logicians in the last forty years. Discussions of the paradoxes of self-reference tend to get very technical very fast. Good introductory discussions can be found in

Read (1994, ch. 6) and Sainsbury (1995, chs. 5, 6). The whole area remains highly contentious.

Chapter 6: Necessity and possibility

The study of inferences involving modal operators goes back to Aristotle, and was continued in the Middle Ages. The modern investigations were started by the US philosopher C. I. Lewis, roughly between 1915 and 1930. The notion of a possible world is to be found in Leibniz, but the way it is applied in this chapter is due largely to another US philosopher, Saul Kripke, who produced the ideas in the 1960s. A standard introduction to the area is Hughes and Cresswell (1996); but you are unlikely to get much out of this before you have mastered an introductory logic book of a more standard kind. Aristotle's argument for fatalism comes from *De Interpretatione*, ch. 9. He thought it fallacious, though not for the reasons given in this chapter. A reasonably accessible discussion of it can be found in Haack (1974, ch. 3). The argument with which the chapter finishes is a version of the 'Master Argument' put forward by the Megarian logician Diodorus Cronus.

Chapter 7: Conditionals

Debate about the nature of conditionals goes back to the Megarians and Stoics, who produced a number of different theories. The issue was also widely discussed in the Middle Ages. The idea that the conditional is truth-functional is one of the Megarian views. It was endorsed in early modern logic by Frege and Russell. The account given in this chapter can certainly be found in medieval logic; in its modern form, it is due to C. I. Lewis, who developed modal logic around it. The notion of conversational implicature is due to the British philosopher Paul Grice in the 1970s (though he used it in defence of the material conditional). The nature of conditionals remains highly contentious. Read (1994, ch. 3) is a readable introduction, as is Part 1 of Sanford (1989).

Chapter 8: The future and the past

Temporal reasoning is discussed by a number of medieval logicians. The approach described in this chapter was invented largely by the New Zealand logician Arthur Prior in the 1960s, inspired by developments in modal logic. A readable account of the subject can be found in Øhrstrøm and Hasle (1995). McTaggart's argument appeared originally in 1908, though his presentation is somewhat different from mine. My presentation follows Mellor (1981, ch. 7).

Chapter 9: Identity and change

Confusion between the *is* of identity and the *is* of predication is a commonplace in the history of philosophy. Though the distinction is drawn by Abelard (1079–1142) and Leibniz (who enunciated 'Leibniz's Law'), it does not really become well understood until the end of the 19th century. There are presentations in most standard modern logic texts, such as Lemmon (1971) and Hodges (1977). Puzzles about identity are legion in philosophy. The one with which the chapter ends is due, as far as I know, to Prior.

Chapter 10: Vagueness

Sorites problems go back to Megarian logic. The problem with which the chapter starts is a version of one called the *Ship of Theseus*, a ship which was, supposedly, rebuilt plank by plank. The example is used first by Plutarch in his *Life of Theseus*, and is later picked up by Thomas Hobbes in the section *De Corpore* of his *Elements of Philosophy*. Intense investigation of problems of this kind is largely a feature of the last fifty years. The logical details described in this chapter were developed initially by the Polish logician Jan Łukasiewicz (pronounced *Woo/ka/zye/vitz*) in the 1920s, quite independently of worries about vagueness. (He was motivated initially by Aristotle's argument about fatalism.) Good discussions of vagueness can be found in Read (1994, ch. 7) and Sainsbury (1995, ch. 2). A much lengthier introduction is Williamson (1994).

Chapter 11: Probability

Historically, inductive validity is quite under-developed, compared with deductive validity. Probability theory was developed in the 18th century, in connection with games of chance, largely by French-speaking mathematicians, such as Pierre de Laplace and members of the prodigious Bernoulli family. The idea of applying it to inductive inference is due mainly to the German logician Rudolf Carnap in the 1950s. There are many notions of probability. The one described in this chapter is usually called the *frequency interpretation*. A good introduction to the whole area is Skyrms (1975).

Chapter 12: Inverse probability

Investigations of the connection between inverse probabilities go back to the 18th-century British mathematician, Thomas Bayes. The connection described in this chapter is often (incorrectly) called Bayes' Theorem. Problems concerning the Principle of Indifference also go back to the origins of probability theory. A standard introduction to reasoning of this kind is Howson and Urbach (1989); but this is not a book for those with a fear of mathematics.

Chapter 13: Decision theory

Decision theory also has its roots in the investigations of probability theory of the 18th century, but became a serious business in the 20th century, with many important applications being found in economics and game theory. A good introduction is Jeffrey (1985), though, again, this book is not for those with a fear of mathematics. The problem with which the chapter ends comes from Gracely (1988).

A number of the arguments we have met in this book concern God, one way or another. This is not because God is a particularly logical topic. It is just that philosophers have had a long time to come up with interesting arguments concerning God. In Chapter 3, we met the Cosmological Argument. Perhaps the most famous version of this was proposed by the medieval philosopher Thomas Aquinas. (His version is much more sophisticated than the argument of Chapter 3, and does not suffer from the problem pointed out there.) The Ontological Argument for the existence of God was proposed by the medieval philosopher Anselm of Canterbury. The version given in Chapter 4 is essentially due to the 17th-century philosopher René Descartes in his *Fifth Meditation*. Biological versions of the Argument to Design were popular in the 19th century, but were destroyed by the Theory of Evolution. Cosmological versions, of the kind given in Chapter 12, became very popular in the 20th century. A good little reference work on arguments for the existence of God is Hick (1964).

Chapter 14: Halt! What goes there?

The theory of computation was worked out by logicians and mathematicians such at Alan Turing, Alonzo Church, and John von Neumann, in the first half of the 20th century, a long time before the existence of computers of the kind with which we are now familiar. This included Turing's proof of the Halting Theorem, and the formulation of the Church-Turing Thesis. For a general discussion of these, see Copeland (2004). The notion of hypercomputation is much more recent. For some discussion, see Piccini (2015). On Turing, his life and his work, see Hodges (2013).

Chapter 15: Maybe it is true—but you can't prove it!

Hilbert's Program was one of several ideas proposed in the first part of the 20th century to provide mathematics with a secure foundation. For discussion of the Program, see Zach (2015). For a much more technical introduction to Gödel's theorems (though not very technical as these things go), see Smith (2007). Gödel's Incompleteness Theorems are, arguably, the most spectacular of his results; however, he proved several other very important results in the foundations of logic and set theory. For a discussion of these, and also on Gödel himself, see Dawson (1997). Gödel's Theorems have been held (rightly or wrongly) to have many philosophical consequences. For a discussion of some of these, see Raatikainen (2005).

* * *

There is, of course, much more to the history of logic than the above details tell. Likewise, there is much to logic itself that is entirely absent from this book. We have been skating over the surface. Chapters 14 and 15 may give the reader some sense of what lies beyond the more basic material in the book; but this is little more than a gesture. Logic is undoubtedly a technical subject; but the roots of its forest of technical ideas and results sink deep into philosophical soil. The great logicians of the past have of course been concerned with the forest; however, most have done so because of an engagement with the philosophical ground. If I have been able to show the reader something of these engagements in the chapters of this book, I can ask no more.

Glossary

The following glossary contains the terms of art and logical symbols that are employed in this book. The entries are not meant to be precise definitions, but are meant to convey the main idea for quick reference. By and large, the terms and symbols are reasonably standard, though there are several other sets of symbols that are also in common use.

algorithm: a procedure that can be carried out in steps where there is no guessing or creativity.

antecedent: what follows the 'if' in a conditional.

arithmetic: the branch of mathematics dealing with the natural numbers (0, 1, 2...).

axiom: the basic statements of an axiom system.

axiom system: a collection of basic statements from which others can be proved by deducing them.

binary numeral: a numeral such as 10011, which expresses a number in terms of powers of 2.

Church–Turing Thesis: The thesis that every algorithm can be performed by a computer program.

code number: a number that can be assigned to an entity such as a statement, computer program, or proof. Given the code number, one can 'decode' to find the thing of which it is the code.

completeness: an axiom system is complete if it can prove every true sentence expressible in its language (and so, given Excluded Middle, either A or $\neg A$, for any A).

conclusion: the part of an inference for which reasons are given.

conditional: if…then….

conditional probability: the probability of some statement, given some other information.

conjunction: … and….

conjuncts: the two sentences involved in a conjunction.

consequent: what follows the 'then' in a conditional.

consistency: an axiom system is consistent if there is no formula, A, such that it can prove both A and $\neg A$.

conversational implicature: an inference, not from what is said, but from the fact that it is said.

decision theory: the theory of how to make decisions under conditions of uncertain information.

deductive validity: an inference is deductively valid when the premisses cannot be true without the conclusion also being true.

(definite) description: a name of the form 'the thing with such and such properties'.

disjunction: either…or….

disjuncts: the two sentences involved in a disjunction.

Excluded Middle: the principle that $A \vee \neg A$, for every A.

expectation: the result of taking each possible outcome, multiplying its value by its probability, and adding all the results together.

fuzzy logic: a kind of logic in which sentences take truth values that may be any number between 0 and 1.

Gödel's (first) Incompleteness Theorem: given a suitably expressive axiomatization of arithmetic, it is either inconsistent or incomplete.

Gödel's (second) Incompleteness Theorem: given a suitably expressive axiomatization of arithmetic, if it is consistent, its consistency cannot be proved in the system.

Halting Theorem: Turing's result that there is no computer program which will determine whether an arbitrary program with an arbitrary input will halt.

Hilbert's Program: the program of axiomatizing all mathematics, and then proving the axiom system consistent.

inductive validity: an inference is inductively valid when the premisses provide some reasonable ground for the conclusion, though not necessarily a conclusive one.

inference: a piece of reasoning, where premisses are given as reasons for a conclusion.

inverse probability: the relationship between the conditional probability of a given b, and of b given a.

'is' of identity: ... is the same object as....

'is' of predication: part of a predicate indicating the application of the property expressed by the rest of it.

Leibniz's Law: if two objects are identical, any property of one is a property of the other.

liar paradox: 'This sentence is false'.

Löb's Theorem: given a suitably expressive axiom system for arithmetic, if it can prove $\exists x Prov(x, \langle A \rangle) \supset A$, it can prove A.

material conditional: not both (... and not ...).

modal operator: a phrase attaching to a sentence, to form another sentence expressing the way in which the first sentence is true or false (possibly, necessarily, etc.).

modern logic: the logical theories and techniques arising out of the revolution in logic around the turn of the 20th century.

modus ponens: the form of inference $a, a \rightarrow c / c$.

name: grammatical category for a word that refers to an object (all being well).

necessity: it must be the case that....

negation: it is not the case that....

particular quantifier: something is such that....

possibility: it may be the case that....

possible world: a situation associated with another, s, where things actually are as they merely might be in s.

predicate: for the grammatically simplest kind of sentence, the part which expresses whatever is said about what the sentence is about.

premisses: the part of an inference that gives reasons.

Principle of Indifference: given a number of possibilities, with no relevant difference between them, they all have the same probability.

prior probability: the probability of some statement before any evidence is taken into account.

probability: a number between 0 and 1, measuring how likely something is.

proof: a deduction in an axiom system.

proper name: a name that is not a description.

quantifier: a word or phrase that can be the subject of a sentence, but which does not refer to an object (*some*..., *all*..., *no*...)

reductio ad absurdum: a method of proof in which one *assumes* the negation of what one wishes to demonstrate, and shows this to be impossible.

reference class: the group of objects from which probability ratios are computed.

Russell's paradox: concerns the set of all sets that are not members of themselves.

self-reference: a sentence or other construction which reflects back on itself.

situation: a state of affairs, maybe hypothetical, in which premisses and conclusions may be true or false.

sorites paradox: a kind of paradox involving repeated applications of a vague predicate.

subject: for the grammatically simplest kind of sentence, the part which tells you what the sentence is about.

syllogism: a form of inference with two premisses and a conclusion, a theory of which was first produced by Aristotle.

tense: past, present, or future.

tense operator: a phrase attaching to a sentence, to form another sentence expressing when the first sentence is true or false (past or future).

theorem: the statements that can be proved in an axiom system.

traditional logic: logical theories and techniques that were employed before the 20th century.

truth conditions: sentences that spell out how the truth value(s) of a sentence depend(s) on the truth values of its components.

truth function: a logical symbol which, when applied to sentences to give a more complex sentence, is such that the truth value of the compound is completely determined by the truth value(s) of its component(s).

truth table: a diagram depicting truth conditions.

truth value: true (T) or false (F).

universal quantifier: everything is such that....

vagueness: a property of a predicate expressing the idea that small changes in an object make no difference to the applicability of the predicate.

valid: applies to an inference in which the premisses really do provide a reason of some kind for the conclusion.

Symbol	Meaning	Name
T	true (in a situation)	truth values
F	false (in a situation)	
$\vee$	either...or	disjunction
&	... and...	conjunction
$\neg$	it's not the case that...	negation
$\exists x$	some object, x, is such that...	particular quantifier
$\forall x$	every object, x, is such that...	universal quantifier
ιx	the object, x, such that...	description operator
$\Box$	it must be the case that...	modal operators
$\Diamond$	it may be the case that...	
$\rightarrow$	if...then...	conditional
$\supset$	not both (...and not...)	material conditional

(*continued*)

Symbol	Meaning	Name
P	it was the case that...	tense operators
F	it will be the case that...	
H	it has always been the case that...	
G	it will always be the case that...	
$=$	...is the same object as...	identity
$<$	...is less than	
$\leq$	...is less than or equal to...	
$\|\ldots\|$	the number which is the truth value of...	
Max	the greater of...and...	
Min	the lesser of...and...	
pr	the probability that...	
$pr(\ldots \| \ldots)$	the probability that...given that...	conditional probablity
E	the expectation of its being the case that...	
V	the value of its being the case that...	
$\simeq$	...is approximately equal to...	
$\langle A \rangle$	The name (code number) of A	code number
Prov(x,y)	x is a proof of y	proof predicate

Problems

For each of the main chapters of the book, the following gives an exercise whereby you can test your understanding of the contents of that chapter. Solutions to the problems can be found in 'Problem Solutions' section, which follows this section.

Chapter 1 Is the following inference deductively valid, inductively valid, or neither? Why? *José is Spanish; most Spanish people are Catholics; so José is Catholic.*

Chapter 2 Symbolize the following inference, and evaluate its validity. *Either Jones is a knave or he is a fool; but he is certainly a knave; so he is not a fool.*

Chapter 3 Symbolize the following inference, and evaluate its validity. *Someone either saw the shooting or heard it; so either someone saw the shooting or someone heard it.*

Chapter 4 Symbolize the following inference, and evaluate its validity. *Everyone wanted to win the prize; so the person who won the race wanted to win the prize.*

Chapter 5 Symbolize the following inference, and evaluate its validity. *You made an omelette, and you don't make an omelette and not break an egg; so you broke an egg.*

Chapter 6 Symbolize the following inference, and evaluate its validity. *It's impossible for pigs to fly, and it's impossible for pigs to breathe under water; so it must be the case that pigs neither fly nor breathe under water.*

Chapter 7 Symbolize the following inference, and evaluate its validity. *If you believe in God, then you go to church; but you go to church; so you believe in God.*

Chapter 8 Symbolize the following inference, and evaluate its validity. *It has always rained, and it always will rain; so it's raining now.*

Chapter 9 Symbolize the following inference, and evaluate its validity. *Pat is a woman, and the person who cleaned the windows is not a woman; so Pat is not the person who cleaned the windows.*

Chapter 10 Symbolize the following inference, and evaluate its validity, where the level of acceptability is 0.5. *Jenny is clever; and either Jenny is not clever or she is beautiful; so Jenny is beautiful.*

Chapter 11 The following set of statistics was collected from ten people (called 1–10).

	1	2	3	4	5	6	7	8	9	10
Tall	✓		✓		✓				✓	
Wealthy	✓		✓		✓		✓	✓		
Happy	✓	✓		✓	✓			✓	✓	

If r is a randomly chosen person in this collection, assess the inductive validity of the following inference: *r is tall and wealthy; so r is happy.*

Chapter 12 Suppose there are two illnesses, A and B, that have exactly the same observable symptoms. Ninety per cent of those who present with the symptoms have illness A; the other 10 per cent have illness B. Suppose, also, that there is a pathology test to distinguish between A and B. The test gives the correct answer nine times out of ten.

1. What is the probability that the test, when applied to a randomly chosen person with the symptoms, will say that they have illness B? (Hint: consider a typical sample of 100 people with the symptoms, and work out how many the test will say to have illness B.)
2. What is the probability that someone with the symptoms has illness B, given that the test says that they do? (Hint: you have to use the first question.)

Chapter 13 You hire a car. If you do not take out insurance, and you have an accident, it will cost you $1,500. If you take out insurance, and have an accident, it will cost you $300. The insurance costs $90, and you estimate that the probability of an accident is 0.05. Assuming that the only considerations are financial ones, should you take out the insurance?

Chapter 14 What is wrong with the following argument? Of course there is an algorithm for determining whether a program with a given input terminates. We simply run the program with that input, and see what happens. Either it will terminate or it will not. Either way, we will have a result.

Chapter 15 Say that an axiom system has the *disjunction property* just if whenever one can prove something of the form $A \vee B$, one can prove either A or B (or both). Suppose that we have a consistent axiom system for arithmetic, all of whose theorems are true, and whose logic is that of Chapter 2. Can it have the disjunction property? (Hint: use the fact that in this logic the law of excluded middle is valid—that is, the logic can prove everything of the form $A \vee \neg A$. Then use Gödel's sentence G.)

Problem solutions

The following are solutions to the problems in the previous section, 'Problems'. In many cases, especially where an inference is invalid, the solutions are not unique: other, equally good, solutions are quite possible.

Chapter 1

Is the following inference deductively valid, inductively valid, or neither? Why?

José is Spanish; most Spanish people are Catholics; so José is Catholic.

The inference is not deductively valid. It is quite possible that the premisses are true, and yet that José is one of the minority of Spanish who are not Catholic. Nonetheless, the premisses together give good (though not decisive) reason for supposing the conclusion to be true. Hence, the inference is inductively valid.

Chapter 2

Symbolize the following inference, and evaluate its validity.

Either Jones is a knave or he is a fool; but he is certainly a knave; so he is not a fool.

Let:

> k be 'Jones is a knave'.
> f be 'Jones is a fool'.

Then the inference is:

$$\frac{k \lor f \quad k}{\neg f}$$

Testing gives:

k	f	$k \lor f$	k	$\neg f$
T	T	T	T	F
T	F	T	T	T
F	T	T	F	F
F	F	F	F	T

On the first row, both premisses are T, and the conclusion is F. Hence, the inference is invalid.

Chapter 3

Symbolize the following inference, and evaluate its validity.

Someone either saw the shooting or heard it; so either someone saw the shooting or someone heard it.

Let:

> xS be 'x saw the shooting'.
> xH be 'x heard the shooting'.

And let the objects in question be people. Then the inference is:

$$\frac{\exists x(xS \vee xH)}{\exists x\, xS \vee \exists x\, xH}$$

This inference is valid. For suppose the premiss is true in some situation. Then there is some object, x, in the domain of that situation such that $xS \vee xH$. By the truth conditions for $\vee$, either xS or xH. In the first case, $\exists x\, xS$; in the second, $\exists x\, xH$. In either case, $\exists x\, xS \vee \exists x\, xH$ is true in the situation.

Chapter 4

Symbolize the following inference, and evaluate its validity.

Everyone wanted to win the prize; so the person who won the race wanted to win the prize.

Let:

xP be 'x wanted to win the prize'.
xR be 'x won the race'.

And let the objects in question be people. Then the inference is:

$$\frac{\forall x\, xP}{(\iota x\, xR)P}$$

The inference is invalid. Take a situation, s, in which everyone satisfies P, but in which no one satisfies R. (Maybe the race was called off!) Then the premiss is true in s. But the description $\iota x\, xR$ does not refer to anything. Hence the conclusion is false in s.

Chapter 5

Symbolize the following inference, and evaluate its validity.

You made an omelette, and you don't make an omelette and not break an egg; so you broke an egg.

Let:

m be 'You made an omelette'.
b be 'You broke an egg'.

Then the inference is:

$$\frac{m \quad \neg(m \,\&\, \neg b)}{b}$$

This inference is invalid. For take the following situation:

b: F but not T
m: T and F

Then $\neg b$ is T (and not F); so $m \,\&\, \neg b$ is T and F (both conjuncts are true, and one is false); so $\neg(m \,\&\, \neg b)$ is T and F. In this situation, both premisses are T, and the conclusion is not.

Chapter 6

Symbolize the following inference, and evaluate its validity.

It's impossible for pigs to fly, and it's impossible for pigs to breathe under water; so it must be the case that pigs neither fly nor breathe under water.

Let:

f be 'Pigs fly'.
b be 'Pigs breathe under water'.

Then the inference is:

$$\frac{\neg\Diamond f \,\&\, \neg\Diamond b}{\Box(\neg f \,\&\, \neg b)}$$

This inference is valid. For suppose the premiss is true in some situation, s. Then both conjuncts are true in that situation. Hence, there is no associated situation, s', where either f is true (first conjunct) or b is true (second conjunct). That is, in every associated situation, s', $\neg f \& \neg b$ is true. Hence, the conclusion is true in s.

Chapter 7

Symbolize the following inference, and evaluate its validity.

If you believe in God, then you go to church; but you go to church; so you believe in God.

Let:

b be 'You believe in God'.
c be 'You go to church'.

Then the inference is:

$$\frac{b \to c \quad c}{b}$$

This inference is invalid. For consider a situation, s, with one associated situation, s', where things are as depicted in the following diagram:

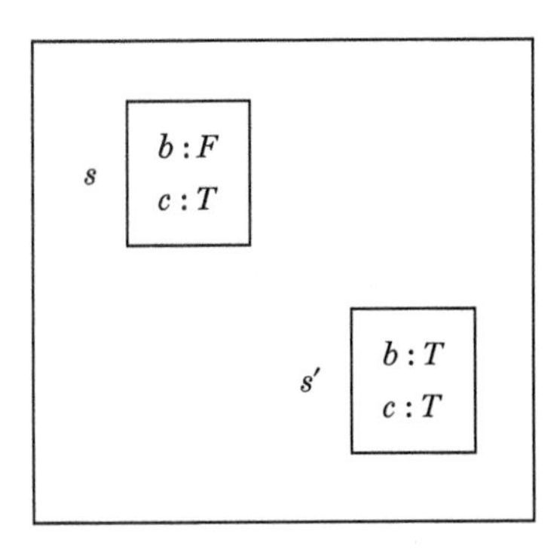

In every situation where b is true, so is c. Hence $b \to c$ is true in s. Thus, both premisses are true in s, but the conclusion is not.

Chapter 8

Symbolize the following inference, and evaluate its validity.

It has always rained, and it always will rain; so it's raining now.

Let:

r be 'It is raining'.

Then the inference is:

$$\frac{\mathbf{H}r \,\&\, \mathbf{G}r}{r}$$

This inference is invalid. For suppose that things are as depicted in the following collection of situations:

$\ldots s_{-3}$	s_{-2}	s_{-1}	s_0	s_1	s_2	$s_3 \ldots$
r	r	r	$\neg r$	r	r	r

r is true at all times before s_0; so $\mathbf{H}r$ is true in s_0. r is true at all times after s_0; so $\mathbf{G}r$ is true in s_0. Hence, $\mathbf{H}r \& \mathbf{G}r$ is true in s_0, but the conclusion is not true in s_0.

Chapter 9

Symbolize the following inference, and evaluate its validity.

Pat is a woman, and the person who cleaned the windows is not a woman; so Pat is not the person who cleaned the windows.

Let:

p be 'Pat'.
c be 'the person who cleaned the windows'.
W be 'is a woman'.

Then the inference is:

$$\frac{pW \,\&\, \neg cW}{\neg p = c}$$

This inference is valid. For take any situation where the premiss is true. Then in that situation, whatever the name p refers to has the property expressed by W, and whatever the name c refers to does not. Hence, by Leibniz's Law, p and c denote different things (assuming that nothing can be both true and false!). That is, $\neg p = c$ is true.

Chapter 10

Symbolize the following inference, and evaluate its validity, where the level of acceptability is 0.5.

Jenny is clever; and either Jenny is not clever or she is beautiful; so Jenny is beautiful.

Let:

c be 'Jenny is clever'.
b be 'Jenny is beautiful'.

Then the inference is:

$$\frac{c \quad \neg c \vee b}{b}$$

This inference is invalid. For take a situation where the truth values of c and b are as follows:

$$c : 0.5$$
$$b : 0.2$$

Then the truth value of $\neg c$ in this situation is 0.5 (1 – 0.5), and so the truth value of $\neg c \vee b$ is also 0.5 ($Max(0.5, 0.2)$). But then both premisses are acceptable (≥0.5), and the conclusion is not.

Chapter 11

The following set of statistics was collected from ten people (called 1–10).

	1	2	3	4	5	6	7	8	9	10
Tall	✓		✓		✓				✓	
Wealthy	✓		✓		✓		✓	✓		
Happy	✓	✓		✓	✓			✓	✓	

If r is a randomly chosen person in this collection, assess the inductive validity of the following inference: *r is tall and wealthy; so r is happy.*

Let:

t be 'r is tall'.
w be 'r is wealthy'.
h be 'r is happy'.

The inference is valid. For there are three people who are tall and wealthy, and two of them are happy. Hence, $pr(h \mid t \,\&\, w) = 2/3$. One of them is unhappy, so $pr(\neg h \mid t \,\&\, w) = 1/3$. Hence, $pr(h \mid t \,\&\, w) > pr(\neg h \mid t \,\&\, w)$.

Chapter 12

Suppose there are two illnesses, A and B, that have exactly the same observable symptoms: 90% of those who present with the symptoms have illness A; the other 10% have illness B. Suppose, also, that there is a pathology test to distinguish between A and B. The test gives the correct answer 9 times out of 10.

1. What is the probability that the test, when applied to a randomly chosen person with the symptoms, will say that they have illness B? (Hint: consider a typical sample of 100 people with the symptoms, and work out how many the test will say to have illness B.)

2. What is the probability that someone with the symptoms has illness B, given that the test says that they do? (Hint: you have to use the first question.)

For Part 1: consider a typical sample of 100 people with the symptoms: 90 will have illness A, and 10 will have illness B. Since the test gives the correct result 9 times out of 10, it will say that 81 of the 90 have A ($90 \times 9/10$), and 9 of them have B. Of the 10 with illness B, it will say that 9 have illness B and 1 has illness A. Hence a total of 18 will be said to have B, and so the probability of a (randomly chosen) person being shown to have B is 18/100.

For Part 2: let r be a randomly chosen person with the symptoms, and let:

b be 'r has illness B'.
t be 'The test says that r has illness B'.

Then:

$pr(t|b) = 9/10$, since the test is 90% accurate;
$pr(b) = 1/10$, since one person in every ten has illness B; and
$pr(t) = 18/100$, by Part 1.

By the relationship between inverse probabilities,

$$pr(b\,|\,t) = pr(t\,|\,b) \times pr(b)\,/\,pr(t) = \frac{9}{10} \times \frac{1}{10} \div \frac{18}{100} = 1/2.$$

Chapter 13

You hire a car. If you do not take out insurance, and you have an accident, it will cost you $1,500. If you take out insurance, and have an accident, it will cost you $300. The insurance costs $90, and you estimate that the probability of an accident is 0.05. Assuming that the only considerations are financial ones, should you take out the insurance?

Tabulate the information as follows:

	Have an accident	Don't have an accident
Take out insurance (t)	0.05\–390	0.95\–90
Don't take out insurance ($\neg t$)	0.05\–1,500	0.950

Calculating expectations, we get:

$$E(t) = 0.05 \times (-390) + 0.95 \times (-90) = -105$$
$$E(\neg t) = 0.05 \times (-1{,}500) + 0.95 \times 0 = -75$$

Since $E(\neg t) > E(t)$, you should not take out insurance.

Chapter 14

What is wrong with the following argument? Of course there is an algorithm for determining whether a program with a given input terminates. We simply run the program with that input, and see what happens. Either it will terminate or it will not. Either way, we will have a result.

We can certainly run the program with the given input. If it does terminate then it will do so sooner or later, and we will then know that it terminates (though we may not know in advance how long it will take to do so). If it does not terminate, however, we will never know this. However long the computation has been going on, if it has not stopped, this may be because it is never going to terminate; but it may just be that it is going to terminate, though we haven't got there yet. There is no way of knowing which of these situations we are in.

Chapter 15

Say that an axiom system has the disjunction property just if whenever one can prove something of the form $A \vee B$, one can prove either A or B (or both). Suppose that we have a consistent axiom system for arithmetic, all of whose theorems are true, and whose logic is that of Chapter 2. Can it have the disjunction property? (Hint: use the fact that in this logic the law of excluded middle is valid—that is, the logic can prove everything of the form $A \vee \neg A$. Then use Gödel's sentence G.)

No. If n is the code of the sentence $\neg\exists x Prov(x, n)$ then since the logic can establish the law of excluded middle, the theory can prove that $\exists x Prov(x, n) \vee \neg\exists x Prov(x, n)$. But Gödel's Theorem shows that $\neg\exists x Prov(x, n)$ cannot be proved, though this is true. But then $\exists x Prov(x, n)$ is false; so it cannot be proved either.

Bibliography

Copeland, B. J. 'Computable Numbers: A Guide', pp. 5–57 of B. J. Copeland (ed.), *The Essential Turing* (Oxford: Oxford Univeristy Press, 2004).

Dawson, John W. *Logical Dilemmas: The Life and Work of Kurt Gödel* (Wellesley, MA: A. K. Peters, Ltd., 1997).

Gracely, E. J. 'Playing Games with Eternity: the Devil's Offer', *Analysis* 48 (1988), p. 113.

Haack, S. *Deviant Logic* (Cambridge: Cambridge University Press, 1974).

Heath, P. 'Nothing', pp. 524–5, Vol. 5, of P. Edwards (ed.), *Encyclopedia of Philosophy* (London: Macmillan, 1967).

Heath, P. *The Philosopher's Alice* (New York, NY: St Martin's Press, 1974).

Hick, J. *Arguments for the Existence of God* (London: Collier-Macmillan Ltd, 1964).

Hodges, A. 'Alan Turing', in E. Zalta (ed.), *Stanford Encyclopedia of Philosophy*, <https://plato.stanford.edu/entries/turing/>, 2013.

Hodges, W. *Logic* (London: Penguin Books, 1977).

Howson, C. and Urbach, P. *Scientific Reasoning: the Bayesean Approach* (La Salle, IL: Open Court, 1989).

Hughes, G. E. and Cresswell, M. *A New Introduction to Modal Logic* (London: Routledge, 1996).

Jeffrey, R. *The Logic of Decision* (Chicago: University of Chicago Press, 2nd edition, 1983).

Kneale, W. and M. *The Development of Logic* (Oxford: Clarendon Press, 1975).

Lemmon, E. J. *Beginning Logic* (London: Thomas Nelson and Sons Ltd, 1971).

Mellor, D. H. *Real Time* (Cambridge: Cambridge University Press, 1981. 2nd edition, London: Routledge, 1998).

Øhrstrøm, P. and Hasle, P. F. V. *Temporal Logic: from Ancient Ideas to Artificial Intelligence* (Dordrecht: Kluwer Academic Publishers, 1995).

Piccini, G. 'Hypercomputation', sec. 4.3 of 'Computation in Physical Systems', in E. Zalta (ed.), *Stanford Encyclopedia of Philosophy*, <https://plato.stanford.edu/entries/computation-physicalsystems/>, 2015.

Priest, G. *Introduction to Non-Classical Logic* (Cambridge: Cambridge University Press, 2nd edition, 2008).

Raatikainen, P. 'On the Philosophical Relevance of Gödel's Incompleteness Theorems', *Revue Internationale de Philosophie* 59 (2005), pp. 513–34.

Read, S. *Thinking about Logic: an Introduction to the Philosophy of Logic* (Oxford: Oxford University Press, 1994).

Sainsbury, R. M. *Paradoxes* (Cambridge: Cambridge University Press, 2nd edition, 1995).

Sanford, D. H. *If P then Q: Conditionals and the Foundations of Reasoning* (London: Routledge, 1989).

Skyrms, B. *Choice and Chance* (Encino, CA: Dickenson Publishing Co., 1975).

Smith, P. *An Introduction to Gödel's Theorems* (Cambridge: Cambridge University Press, 2007).

Strawson, P. *Introduction to Logical Theory* (London: Methuen & Co., 1952).

Williamson, T. *Vagueness* (London: Routledge, 1994).

Zach, R. 'Hilbert's Program', in E. Zalta (ed.), *Stanford Encyclopedia of Philosophy*, <https://plato.stanford.edu/entries/hilbert-program/>, 2015.

Zalta, E. *Stanford Encyclopedia of Philosophy*, <https://plato.stanford.edu>, 1995–.

General index

G

H

I

L

M

N

O

P

Q

R

S

T

V

Index of names

BEAUTY

A Very Short Introduction

Roger Scruton

In this *Very Short Introduction* the renowned philosopher Roger Scruton explores the concept of beauty, asking what makes an object - either in art, in nature, or the human form - beautiful, and examining how we can compare differing judgements of beauty when it is evident all around us that our tastes vary so widely. Is there a right judgement to be made about beauty? Is it right to say there is more beauty in a classical temple than a concrete office block, more in a Rembrandt than in last year's Turner Prize winner? Forthright and thought-provoking, and as accessible as it is intellectually rigorous, this introduction to the philosophy of beauty draws conclusions that some may find controversial, but, as Scruton shows, help us to find greater sense of meaning in the beautiful objects that fill our lives.

A fascinating book, which I heartily recommend.

Brya Wilson, Readers Digest

www.oup.com/vsi

Critical Theory

A Very Short Introduction

Stephen Eric Bronner

In its essence, Critical Theory is Western Marxist thought with the emphasis moved from the liberation of the working class to broader issues of individual agency. Critical Theory emerged in the 1920s from the work of the Frankfurt School, the circle of German-Jewish academics who sought to diagnose--and, if at all possible, cure--the ills of society, particularly fascism and capitalism. In this book, Stephen Eric Bronner provides sketches of famous and less famous representatives of the critical tradition (such as George Lukács and Ernst Bloch, Theodor Adorno and Walter Benjamin, Herbert Marcuse and Jurgen Habermas) as well as many of its seminal texts and empirical investigations.

German Philosophy

A Very Short Introduction

Andrew Bowie

German Philosophy: A Very Short Introduction discusses the idea that German philosophy forms one of the most revealing responses to the problems of 'modernity'. The rise of the modern natural sciences and the related decline of religion raises a series of questions, which recur throughout German philosophy, concerning the relationships between knowledge and faith, reason and emotion, and scientific, ethical, and artistic ways of seeing the world. There are also many significant philosophers who are generally neglected in most existing English-language treatments of German philosophy, which tend to concentrate on the canonical figures. This *Very Short Introduction* will include reference to these thinkers and suggests how they can be used to question more familiar German philosophical thought.

www.oup.com/vsi